PLAYING WITH INFINITY

MATHEMATICAL EXPLORATIONS AND EXCURSIONS

BY

PROFESSOR RÓZSA PÉTER

Eötvös Loránd University, Budapest

TRANSLATED BY

DR. Z. P. DIENES

DOVER PUBLICATIONS, INC.
NEW YORK

Dedicated to my brother, Dr. Nicolas Politzer,
who perished at Colditz in Saxony, 1945

Published in Canada by General Publishing
Company, Ltd., 30 Lesmill Road, Don Mills,
Toronto, Ontario.
Published in the United Kingdom by Constable
and Company, Ltd., 10 Orange Street, London
WC 2.

This Dover edition, first published in 1976, is
an unabridged and unaltered republication of the
English translation first published in England in
1961. It is republished through special arrangement with the original publisher, G. Bell & Sons,
Ltd., York House, Portugal Street, London,
WC2A 2HL, England.
The work was originally published in Hungarian
under the title *Játek a Végtelennel-Matematika
Kívülállóknak* by Bibliotheca Kiadó in Budapest
in 1957.

International Standard Book Number: 0-486-23265-4
Library of Congress Catalog Card Number: 75-26467

Manufactured in the United States of America
Dover Publications, Inc.
180 Varick Street
New York, N.Y. 10014

PREFACE

THIS book is written for intellectually minded people who are not mathematicians. It is written for men of literature, of art, of the humanities. I have received a great deal from the arts and I would now like in my turn to present mathematics and let everyone see that mathematics and the arts are not so different from each other. I love mathematics not only for its technical applications, but principally because it is beautiful; because man has breathed his spirit of play into it, and because it has given him his greatest game—the encompassing of the infinite. Mathematics can give to the world such worthwhile things—about ideas, about infinity; and yet how essentially human it is—unlike the dull multiplication table, it bears on it for ever the stamp of man's handiwork.

The popular nature of the book does not mean that the subject is approached superficially. I have endeavoured to present concepts with complete clarity and purity so that some new light may have been thrown on the subject even for mathematicians and certainly for teachers. What has been left out is the systematization which can so easily become boring; in other words, only technicalities have been omitted. (It is not the purpose of the book to teach anyone mathematical techniques.) If an interested pupil picks up this book it will give him a picture of the whole of mathematics. In the beginning I did not mean the book to be so full; the material expanded itself as I was writing it and the number of subjects which could be omitted rapidly decreased. If there were parts to which memories of boredom previously attached I felt that I was picking up some old piece of furniture and blowing the dust off in order to make it shine.

It is possible that the reader may find the style a little naïve in places, but I do not mind this. A naïve point of view in relation to simple facts always conjures up the excitement of new discovery.

I shall tell the reader in the Introduction how the book originated. The writer of whom I speak there is Marcell

Benedek. I began by writing to him about differentiation and it was his idea that a book could grow out of these letters.

I do not refer to any sources. I have learned a lot from others but today I can no longer say with certainty from whence each piece came. There was no book in front of me while I was writing. Here and there certain similes came to my mind with compelling force, the origins of which I could sometimes remember; for example, the beautiful book by Rademacher and Toeplitz,* or the excellent introduction to analysis by Beke.† Once a method had been formed in my mind I could not really write it in any other way just to be more original. I chiefly refer, in this connexion, to the ideas I gained from László Kalmár. He was a contemporary of mine as well as my teacher in mathematics. Anything I write is inseparably linked with his thoughts. I must mention, in particular, that the 'chocolate example', with the aid of which infinite series are discussed, originated with him, as well as the whole idea of the building up of logarithm tables.

I shall have to quote my little collaborators in the schoolroom by their christian names; they will surely recognize themselves. Here I must mention my pupil Kató, who has just finished the fourth year at the grammar school and contributed to the book while it was being written. It is to her that I must be grateful for being able to see the material with the eyes of a gifted pupil.

The most important help I received was from those people who have no mathematical interests. My dear friend Béla Lay, theatrical producer, who had always believed that he had no mathematical sense, followed all the chapters as they were being written; I considered a chapter finished only when he was satisfied with it. Without him the book perhaps would never have been written.

Pál Csillag examined the manuscript from the point of view of the mathematician; also László Kalmár found time, at the last minute, for a quick look. I am grateful to them for the certainty I feel that everything in the book is right.

Budapest
Autumn, 1943   RÓZSA PÉTER

---

* Rademacher and Toeplitz: *The Enjoyment of Mathematics.*
† Manó Beke: *Introduction to the Differential and Integral Calculus.*
[I mention this here for those who might be eager to follow it up.]

PREFACE TO THE ENGLISH EDITION

SINCE 1943, seventeen eventful years have passed. During this time my mathematician friend, Pál Csillag, and my pupil, Kató (Kató Fuchs), have fallen victims of Fascism. The father of my pupil Anna, who suffered imprisonment for seventeen years for illegal working-class activity, has been freed. In this way perhaps even in Anna's imagination the straight lines forever approaching one another will meet. (See page 218.) No book could appear during the German occupation; a lot of existing copies were destroyed by bombing, the remaining copies appeared in 1945—on the first free book-day.

I am very grateful to Dr. Emma Barton, who took up the matter of the English publication of my book, to Professor Dr. R. L. Goodstein, who brought it to a head, to Dr. Z. P. Dienes for the careful translation and to Messrs. G. Bell & Sons for making possible the propagation of the book in the English-speaking world.

The reader should remember that the book mirrors my methods of thinking as they were in 1943; I have hardly altered anything in it. Only the end has been altered substantially. Since then, László Kalmár and I have proved that the existence of absolutely undecidable problems follows from Gödel's Theorem on relatively undecidable problems, but of course in no circumstances can a consequence be more important than the Theorem from which it follows.

Budapest
1960   RÓZSA PÉTER

CONTENTS

Page

PART III THE SELF-CRITIQUE OF PURE REASON

INTRODUCTION

A CONVERSATION I had a long time ago comes into my mind. One of our writers, a dear friend of mine, was complaining to me that he felt his education had been neglected in one important aspect, namely he did not know any mathematics. He felt this lack while working on his own ground, while writing. He still remembered the co-ordinate system from his school mathematics, and he had already used this in similes and imagery. He felt that there must be a great deal more such usable material in mathematics, and that his ability to express himself was all the poorer for his not being able to draw from this rich source. But it was all, so he thought, quite hopeless, as he was convinced of one thing: he could never penetrate right into the heart of mathematics.

I have often remembered this conversation; it has always suggested avenues of thought to me and plans. I saw immediately that there was something to do here, since in mathematics for me the element of atmosphere had always been the main factor, and this was surely a common source from which the writer and the artist could both draw. I remember an example from my schooldays: some fellow students and I were reading one of Shaw's plays. We reached the point where the hero asked the heroine what was her secret by means of which she was able to win over and lead the most unmanageable people. The heroine thought for a moment and then suggested that perhaps it could be explained by the fact that she really kept her distance from everyone. At this point the student who was reading the part suddenly exclaimed: 'That is just the same as the mathematical theorem we learnt today!' The mathematical question had been: Is it possible to approach a set of points from an external point in such a way that every point of the set is approached simultaneously? The answer is yes, provided that the external point is far enough away from the whole set:

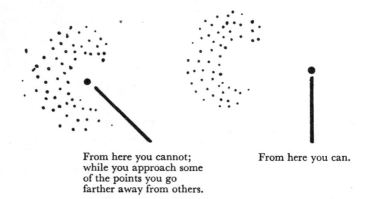

From here you cannot;   From here you can.
while you approach some
of the points you go
farther away from others.

I did not wish to believe the writer's other statement, namely
that he could never penetrate right into the heart of mathematics, that for instance he would never be able to understand
the notion of the differential coefficient. I tried to analyse
the introduction of this notion into the simplest possible,
obvious steps. The result was very surprising; the mathematician cannot even imagine what difficulties the simplest
formula can present to the layman. Just as the teacher cannot
understand how it is possible that a child can spell c–a–t
twenty times, and still not see that it is really a *cat*; and there
is more to this than to a cat!

This again was an experience that caused me to do a great
deal of thinking. I had always believed that the reason why
the public was so ill-informed about mathematics was simply
that nobody had written a good popular book for the general
public about, say, the differential calculus. The interest
patently exists, as the public snaps up everything of this kind
that is available to it; but no professional mathematician has
so far written such a book. I am thinking of the real professional who knows exactly to what extent things can be
simplified without falsifying them, who knows that it is not
a question of serving up the usual bitter pill in a pleasanter
dish (since mathematics for most is a bitter memory); one
who can clarify the essential points so that they hit the eye,
and who himself knows the joy of mathematical creation and
writes with such a swing that he carries the reader along with

him. I am now beginning to believe that for a lot of people even the really popular book is going to remain inaccessible.

Perhaps it is the decisive characteristic of the mathematician that he accepts the bitterness inherent in the path he is travelling. 'There is no royal road to mathematics', Euclid said to an interested potentate; it cannot be made comfortable even for kings. You cannot read mathematics superficially; the inescapable abstraction always has an element of self-torture in it, and the one to whom this self-torture is joy is the mathematician. Even the simplest popular book can be followed only by those who undertake this task to a certain extent, by those who undertake to examine painstakingly the details inherent in a formula until it becomes clear to them.

I am not going to write for these people. I am going to write mathematics without formulae. I want to pass on something of the feel of mathematics. I do not know if such an undertaking can succeed. By giving up the formula, I give up an essential mathematical tool. The writer and the mathematician alike realize that form is essential. Try to imagine how you could express the feel of a sonnet without the form of the sonnet. But I still intend to try. It is possible that, even so, some of the spirit of real mathematics can be saved.

One way of making things easy I cannot allow; the reader must not omit, leave for later reading, or superficially skim through, any of the chapters. Mathematics can be built up only brick by brick; here not one step is unnecessary, for each successive part is built on the previous one, even if this is not quite as obviously so as in a boringly systematic book. The few instructions must be carried out, the figures must really be studied, simple drawings or calculations must really be attempted when the reader is asked to do so. On the other hand I can promise the reader that he will not be bored.

I shall not make use of any of the usual school mathematics. I shall begin with counting and I shall reach the most recent branch of mathematics, mathematical logic.

PART I

THE SORCERER'S APPRENTICE

1. Playing with fingers

LET us begin at the beginning. I am not writing a history of mathematics; this could be done only on the basis of written evidence, and how far from the beginning is the first written evidence! We must imagine primitive man in his primitive surroundings, as he begins to count. In these imaginings, the little primitive man, who grows into an educated human being before our eyes, will always come to our aid; the little baby, who is getting to know his own body and the world, is playing with his tiny fingers. It is possible that the words 'one', 'two', 'three' and 'four' are mere abbreviations for 'this little piggie went to market', 'this little piggie stayed at home', 'this one had roast beef', 'this one had none' and so on; and this is not even meant to be a joke: I heard from a medical man that there are people suffering from certain brain injuries who cannot tell one finger from another, and with such an injury the ability to count invariably disappears. This connexion, although unconscious, is therefore still extremely close even in educated persons. I am inclined to believe that one of the origins of mathematics is man's playful nature, and for this reason mathematics is not only a Science, but to at least the same extent also an Art.

We imagine that counting was already a purposeful activity in the beginning. Perhaps primitive man wanted to keep track of his property by counting how many skins he had. But it is also conceivable that counting was some kind of magic rite, since even today compulsion-neurotics use counting as a magic prescription by means of which they regulate certain forbidden thoughts; for example, they must count from one to twenty and only then can they think of something else. However this may be, whether it concerns animal skins or successive time-intervals, counting always means that we go beyond what is there by one: we can even go beyond our ten fingers and so emerges man's first magnificent mathematical creation, the infinite sequence of numbers,

$$1, 2, 3, 4, 5, 6, \ldots$$

the sequence of natural numbers. It is infinite, because after any number, however large, you can always count one more. This creation required a highly developed ability for abstraction, since these numbers are mere shadows of reality. For example, 3 here does not mean 3 fingers, 3 apples or 3 heartbeats, etc., but something which is common to all these, something that has been abstracted from them, namely their number. The very large numbers were not even abstracted from reality, since no one has ever seen a billion apples, nobody has ever counted a billion heartbeats; we imagine these numbers on the analogy of the small numbers which do have a basis of reality: in imagination one could go on and on, counting beyond any so-far known number.

Man is never tired of counting. If nothing else, the joy of repetition carries him along. Poets are well aware of this; the repeated return to the same rhythm, to the same sound pattern. This is a very live business; small children do not get bored with the same game; the fossilized grown-up will soon find it a nuisance to keep on throwing the ball, while the child would go on throwing it again and again.

We go as far as 4? Let us count one more, then one more, then one more! Where have we got to? To 7, the same number that we should have got to if we had straight away counted 3 more. We have discovered addition

$$4 + 1 + 1 + 1 = 4 + 3 = 7$$

Now let us play about further with this operation: let us add to 3 another 3, then another 3, then another 3! Here we have added 3's four times, which we can state briefly as: four threes are twelve, or in symbols:

$$3 + 3 + 3 + 3 = 4 \times 3 = 12$$

and this is multiplication.

We may so enjoy this game of repetition that it might seem difficult to stop. We can play with multiplication in the same way: let us multiply 4 by 4 and again by 4, then we shall get

$$4 \times 4 \times 4 = 64$$

This repetition or 'iteration' of multiplication is called raising to a power. We say that 4 is the base, and we indicate by means of a small number written at the top right-hand corner of the 4 the number of 4's that we have to multiply; i.e. the notation is this:

$$4^3 = 4 \times 4 \times 4 = 64$$

As is easily seen, we keep getting larger and larger numbers: 4×3 is more than $4 + 3$, and 4^3 is a good deal more than 4×3. This playful repetition carries us well up amongst the large numbers; even more so, if we iterate raising to a power itself. Let us raise 4 to the power which is the fourth power of four:

$$4^4 = 4 \times 4 \times 4 \times 4 = 64 \times 4 = 256$$

and we have to raise 4 to this power:

$$4^{4^4} = 4^{256} = 4 \times 4 \times 4 \times 4 \ldots$$

I have no patience to write any more, since I should have to put down 256 4's, not to mention the actual carrying out of the multiplication! The result would be an unimaginably large number, so that we use our common sense, and, however amusing it would be to iterate again and again, we do not include the iteration of powers among our accepted operations.

Perhaps the truth of the matter is this: the human spirit is willing to play any kind of game that comes to hand, but only those of these mathematical games become permanent features that common sense decides are going to be useful.

Addition, multiplication and raising to powers have proved very useful in man's common-sense activities and so they have gained permanent civil rights in mathematics. We have determined all those of their properties which make calculations easier; for example, it is a great saving that 7×28 can be calculated not only by adding 28 7 times, but also by splitting it into two multiplication processes: 7×20 as well as 7×8 can quite easily be calculated and then it is readily determined how much $140 + 56$ will be. Also in adding long columns of numbers how useful it is to know that no amount of rearranging of the order of the additions is going to spoil the result, as for example $8 + 7 + 2$ can be carried out as $8 + 2 = 10$ and to 10 it is quite easy to add 7; in this way I have cunningly avoided the awkward addition $8 + 7$. We merely have to consider that addition really means counting on by just as much as the numbers to be added and then it becomes clear that changing the order does not alter the result. To be convinced of the same thing about multiplication is a little harder, since 4×3 means $3 + 3 + 3 + 3$ and 3×4 means $4 + 4 + 4$ and it is really not obvious that

$$3 + 3 + 3 + 3 = 4 + 4 + 4$$

But this straightaway becomes clear if we do a little drawing. Let us draw four times three dots in these positions $\quad . \quad . \quad .$ one underneath the other

$$
\begin{matrix}
. & . & . \\
. & . & . \\
. & . & . \\
. & . & .
\end{matrix}
$$

Everyone can see that this is the same thing as if we had drawn three times four dots in the following positions

$$
\begin{matrix}
. \\
. \\
. \\
.
\end{matrix}
$$

next to each other. In this way $4 \times 3 = 3 \times 4$. This is why mathematicians have a common name for the multiplier and the multiplicand; the factors.

Let us look at one of the rules for raising to powers:

$$4 \times 4 \times 4 \times 4 \times 4 = 4^5$$

If we get tired of all this multiplying, we can have a little rest; the product of the first three 4's is 4^3, there is still 4^2 left, so

$$4^3 \times 4^2 = 4^5$$

The exponent of the result is 5, which is $3 + 2$; so we can multiply the two powers of 4 by adding their exponents. As a matter of fact this is always so. For example:

$$5^4 \times 5^2 \times 5^3 = \underbrace{5 \times 5 \times 5 \times 5}_{} \times \underbrace{5 \times 5}_{} \times \underbrace{5 \times 5 \times 5}_{} = 5^9$$

here again $9 = 4 + 2 + 3$.

Let us recapitulate the ground we have covered: it was counting that led to the four rules. It could be objected, where does subtraction come in all this? And division? But these are merely reversals of the operations we have had so far (as are extraction of roots and logarithms). Because, for example, $20 \div 5$ involves our knowing the result of a multiplication sum, namely 20, we are seeking the number which if multiplied by 5 gives 20 as the result. In this case we succeed in finding such a number, since $5 \times 4 = 20$. But it is not always easy to find such numbers; in fact it is not even certain that there is one. For example 5 does not go into 23 without a remainder since $4 \times 5 = 20$ is too little and $5 \times 5 = 25$ is more than 23, and so we are forced to be satisfied with the smaller one and to say that 5 goes 4 times into 23, but 3 is left over. This kind of thing certainly causes more headaches than our playful iterations; the reversed operations are usually bitter operations. It is for this reason that they are favourite points of attack in mathematical research, since mathematicians are well known to take delight in difficulties. So I shall have to return to these reversed or inverse operations in what follows.

2. *The 'temperature charts' of the operations*

We saw that iteration of operations carried us higher and higher in the realm of large numbers.   It is worth spending a little time thinking about just what heights we have reached.

For example, we must raise to a power when we want to calculate the volume of a cube.   We choose some small cube as a unit and the question is how many of these small cubes would fill up a bigger cube.   Let us take for example an inch cube as our unit, i.e. a cube whose length, breadth and height are all one inch.

Let us put four of these little cubes next to each other, and we get a row like this:

Then, if we put four such rows next to each other, we make a layer like this:

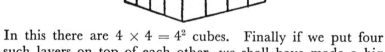

In this there are $4 \times 4 = 4^2$ cubes.   Finally if we put four such layers on top of each other, we shall have made a big cube like this:

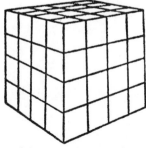

and this is made up of $4 \times 4 \times 4 = 4^3 = 64$ little cubes.

Taking it the other way round, if we start off with the big cube whose length, breadth and height are 4 inches, this can be made up of 4^3 inch cubes; in general, we get the volume of a cube by raising an edge to the third power. This is why we call raising to the third power cubing.*

One consequence of this cubing is that a cube with a relatively short edge will have an enormous volume. For example 1000 yards is not a great distance; everyone can visualize it, if they recall for example that Charing Cross Road in London is about that long. But if we built a cube so that each of its edges was as long as Charing Cross Road, then its volume would be so large that practically the whole of the human race could be accommodated in it. If anyone does not believe this he can make the following calculations: There are no people taller than say $2\frac{1}{2}$ yards (7' 6"), so at every $2\frac{1}{2}$ yards we should make a floor, and so there would be 400 floors to a height of 1000 yards. If we subdivide these floors length-wise as well as across into strips a yard wide, like this

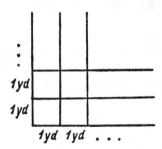

then in every strip we shall have made 1000 squares and there will be 1000 such strips, i.e. there will be $1000 \times 1000 = 1,000,000$ squares on each floor. The length and the breadth of each square is one yard, we can certainly place five people on each of these squares and so we can squeeze in 1,000,000 times 5, i.e. 5 million people quite well into one of these floors. On the 400 floors there will be 400 times 5 million people, i.e. 2,000,000,000, and this is about as many as there are people

* I know the teacher's objection: I should have said that I get the measure of the volume if I raise to the third power the *measure* of an edge.   But I am not going to bore my readers with such hairsplitting.   There is a weightier matter to con-sider: the question is, is it possible to express the edges of any cube in inches?   I shall come back to this later.

in the world, or at least there were not more than this when I
was told about this cube before 1943.

And yet in the calculation of the volume of a cube only the
third power comes in; a larger exponent carries us much faster
still up amongst the large numbers. This fact must have been
a great surprise to the potentate from whom the inventor of
the game of chess modestly asked for only a few grains of wheat
as his reward; he asked for the following to be put on his
chessboard of 64 squares, one grain on the first square, twice

as many on the second square, i.e. 2; twice as many as that on
the third square, i.e. $2 \times 2 = 2^2 = 4$, and so on. At first
this request seems modest enough, but as we run through the
squares, we come across higher and higher powers of 2, until
finally we are dealing with

$$1 + 2 + 2^2 + 2^3 + 2^4 + \ldots + 2^{63}$$

grains of wheat (please imagine that all the powers in between
are there as well; I could not be bothered to write in all the 64
terms), and, if someone cares to work out how much this is,
he will get so much wheat as a result that the whole surface of
the earth could be covered with a half-inch layer of it.

After all this it is not surprising that the iteration of powers
carries us up to such enormous heights. I shall mention just
one fact as a point of interest: it is possible to estimate that 9^{9^9}
is such a large number that just for writing it down you would
need 11,000 miles of paper (writing five digits in every inch),
and a whole lifetime would not be sufficient for its exact
calculation.

As I read over what I have written so far, it strikes me that I have been making use of expressions like 'carries us high up' amongst the numbers, whereas the number series

$$1, 2, 3, 4, 5, \ldots$$

is a horizontal series; by right, I should be able to say only that I am going to the right or at the most that I am going forward towards the large numbers.   The choice of this particular expression must have been influenced by the element of atmosphere: to become larger and larger means to grow, and growth gives rise in us to a feeling of breaking through to new heights.   The mathematician puts this feeling into concrete form: he often accompanies his imaginings with drawings, and the drawing for very rapid growth is the line that rises steeply upwards.

The sick are very familiar with such drawings; they know that they need only to glance at their temperature charts and this shows the whole progress of their illness.   Let us suppose that the following were the temperatures, taken at regular intervals:

$$101, 102, 103, 103, 101, 102, 99, 98$$

These are represented in the following way: first we draw a horizontal line and on this we show the equal time-intervals by equal distances,

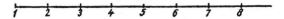

then we choose a certain distance to stand for a degree and from each point of time we draw upwards* that multiple of this distance (corresponding to the rising temperature) which is the sick person's temperature at the point of time in question. But there is no need to draw such long lines, since the temperature never falls below 97, and so we can agree that the height of our horizontal line should correspond to 97.   We shall have to draw above this line one after the other

$$4, 5, 6, 6, 4, 5, 2, 1$$

---

* 'Upwards' as a matter of fact is figurative speech even here, since you can only draw horizontal lines on a piece of paper lying flat.   We still feel that a line in such a direction | points upwards.

degrees.   In this way we get the following picture

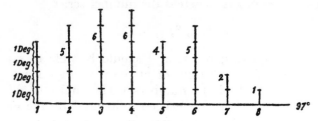

and if we join the end-points of the lines we have drawn, we get

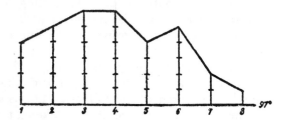

The temperature chart so obtained explains everything.   The
rising lines indicate the rising of the temperature, the level
line shows a stationary period of the illness; the rise in the
beginning was steady, this is shown by the fact that the first
two joins are equally steep, and so form one straight line;
apart from a slight relapse at the sixth time the temperature
was taken, the patient improved rapidly: the fall of the line
joining the 6th and 7th points is very steep, steeper than any
rise.

There is no reason why we should not draw the 'temperature
charts' of our arithmetical operations.

The numbers themselves are usually represented in a similar
way along a line: on this line we pick out an arbitrary starting
point which we call zero and from this point we measure off
equal distances next to one another, i.e. we count in terms of
such distances

Anyone who is handy at counting can carry out the operations
mechanically on such a line: for example, if we were considering the operation 2 + 3, we need only take 3 steps to the right

from the 2 and we can read off the result as being 5.   If we were considering 5 — 3 we should take 3 steps to the left from the 5 and so on.

In a similar way is the abacus used, on the wires of which beads can be moved up or down.

But let us leave the horizontal and go upwards.   Let us start with a certain number, say 3, and let us see how it grows if we add to it 1, then 2, then 3 and so on, or else if we multiply it by 1, then by 2, then by 3, and finally if we raise it to the first, second and third powers ('raise' to a power: in this expression, too, we have the idea of pointing upwards).

Let us begin with addition.   One of the terms is always 3, the other variable term will be represented on the horizontal line and the corresponding sum will point upwards

$$3 + 1 = 4$$
$$3 + 2 = 5$$
$$3 + 3 = 6$$
$$3 + 4 = 7$$

Thus if we represent 1 horizontally by a distance such as this: ⊢———⊣ and vertically by a distance such as this: I the 'temperature chart' for addition will be the following:

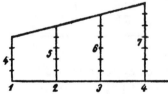

Here every joining line falls onto one and the same straight line. The sum grows steadily as we increase one of its terms.

In the case of multiplication we have:

$$3 \times 1 = 3$$
$$3 \times 2 = 6$$
$$3 \times 3 = 9$$
$$3 \times 4 = 12$$

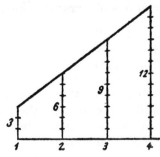

It can be seen that the product also grows steadily if we increase one of its factors, but much more rapidly than the sum: the straight line we get here is a good deal steeper.

Finally if we take powers we have:

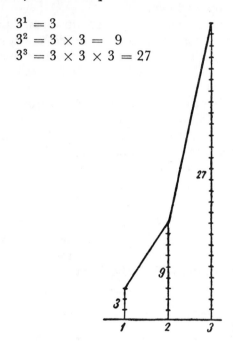

$$3^1 = 3$$
$$3^2 = 3 \times 3 = 9$$
$$3^3 = 3 \times 3 \times 3 = 27$$

The powers do not even grow steadily, but more and more rapidly. There would not even be room for 3^4 on this page. This is the origin of the saying that a certain effect 'increases exponentially'.

In the same way we can construct the charts for the inverse operations; for example for subtraction we should have:

$$3 - 1 = 2$$
$$3 - 2 = 1$$
$$3 - 3 = 0$$

which gives a falling straight line; so the difference decreases steadily if we increase the term to be subtracted.

Division is rather a delicate operation; I shall return to its chart at a later stage.

I shall just make one more remark; what we have been engaged in doing here is what the mathematicians call the graphical representation of functions. The sum depends on our choice of value for the variable term; we express this by saying that the sum is a function of the variable term, and we have represented the growth of this function. In the same way the product is a function of its variable factor, the power of its exponent and so on. Already at the very first operations we have come face to face with functions, and in what follows we shall be examining functional relationships. The notion of function is the backbone of the whole structure of mathematics.

3. *The parcelling out of the infinite number series*

WHAT a long way we have travelled from our games with our fingers! If we have practically forgotten that we have 10 fingers it is only because I did not want to tire my reader with a lot of calculation. Otherwise he would already have noticed that however large a number we write down, we make use of only 10 different symbols, namely,

$$0, 1, 2, 3, 4, 5, 6, 7, 8, 9$$

How is it possible to write down any one of the numbers from the infinite number series by using a mere 10 symbols? It is done by parcelling out this indefinitely increasing number series, by enclosing some of its parts: when we have counted 10 units, we say that we can still grasp that amount at a glance. Let us gather them up in one bundle and call such a bundle a ten, the collective name for the 10 units. We can exchange 10 silver shillings for a single 10-shilling note. Now we can count on in longer steps, progressing by tens; then we can bundle together ten tens, for example we could tie a ribbon round them on which we can write '1 hundred'. Going on like this, we can bundle 10 hundreds into one thousand, 10 thousands into a tenthousand, 10 tenthousands into a hundred-thousand, and 10 hundredthousands into a million. In this way every number can really be written down with the aid of the above-mentioned ten symbols. When we get beyond 9, we write a 1 again, indicating 1 ten. The number after this consists of one ten and one unit, i.e. it can be written down with the aid of two 1's. On the other hand, while writing them down we also have to use the words 'tens', 'hundreds' and similar words. A clever idea makes even these unnecessary: the shop-keeper puts his shilling, two-shilling, half-crown pieces into different sections of his till, the small change on the right because he needs to deal with that a lot in giving change, and towards the left he will put larger and larger denominations. The shopkeeper's hands get so used to this arrange-ment that he will know without looking what kind of coin he is picking up, for example, from the third section. In the same way we could agree about the places in which to put the ones, tens, hundreds. Let us write the ones on the right, and then the larger and larger units, moving along towards the left: the second place is for the tens, the third one for the hundreds. In this way we can leave out the words, because we can recognize the values of the number symbols from their positions; the symbols have thus place-value.

354

consists of 3 hundreds, 5 tens and 4 ones. This is what we mean when we say that we use the decimal system.

On the other hand there would have been no reason why we should not have stopped before or after 10. I have heard of primitive tribes whose knowledge of counting consists of 1, 2, many. We could build a number system even for them: let us bundle the numbers together in twos. The 2 therefore is a new unit, a two, 2 twos again another unit, a four, 2 fours are eight, and so on. In this number system the symbols

0, 1

are sufficient for writing down any number. We can see this most easily in this way: let us suppose that we have coins like this

in other words the units of the binary number system figure as coins. How could we make up 11s. with the smallest possible number of coins? Clearly out of the following three

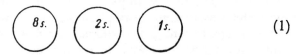 (1)

which give 11s., and out of fewer coins you could never make up 11s. Similarly

 (2)

make 9s., and

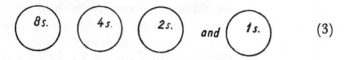

make up 15*s*. The reader should try himself to see that every
number from 1 to 15 can be made up out of

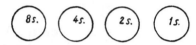

in such a way that each coin is not used more than once, i.e.
either 0 times or 1 times. (We cannot make up 16 in this way
but no wonder, since $2 \times 8 = 16$; a 'sixteen' is in fact the next
unit). According to example (1) 11 can be written in the
binary system as

$$1011$$

since this really means 1 one, 1 two, 0 four, and 1 eight, and
these together do in fact give 11. Similarly we can gather
from examples (2) and (3) that 9 and 15 can be written in the
following ways respectively in the binary system

$$1001, \ 1111$$

So we can really make do with two symbols. It is worth
while having some practice the other way round:
   In the binary system $11101 = 1$ one, 1 four, 1 eight and 1
sixteen $= 1 + 4 + 8 + 16 = 29$ in the decimal system.
   What is the use of a number system? Every operation
becomes unquestionably simpler if we keep the number system
tidy in this way and, for example in additions, we add ones to
ones and tens to tens. The shopkeeper does not add up his
takings in a topsy-turvy way, he counts the like coins in each
section separately and then adds up these sums. Convenience
is an oft-recurring and important factor in the development of
mathematics. The most inconvenient operation is division;
probably all the bother that went with it gave the first push
towards the parcelling out of the number series. How pleasant
are those divisions which can be carried out without remainders!
There are some good, friendly numbers into which a great

number of other numbers will go without remainder. Such a
one is for example 60

$$60 = \begin{bmatrix} 1 \times 60 \\ 2 \times 30 \\ 3 \times 20 \\ 4 \times 15 \\ 5 \times 12 \\ 6 \times 10 \end{bmatrix}$$

and so 1, 2, 3, 4, 5, 6, 10, 12, 15, 20, 30, 60 will all go into 60
without remainder. Thus if we want to divide by one of
these twelve numbers (although it is a pity to include 1 amongst
them as this fortunately has no effect in either multiplication or
division) let us remember that we reached the number we wish
to divide (as every other number) by counting in 1's. Let us
now count 60 of these 1's, then another 60 and so on as far as
we can; the division of these 60's is child's play, and no more
than 59 can be left over, i.e. not a very large number. It is not
much trouble to divide this number, even if there is going to be
a remainder. From this point of view we ought to bundle our
numbers in 60's, and the ancients did in fact introduce the 60
number system for the measurement of angles and time in
connexion with their astronomical activities, which required a
lot of awkward divisions. To this day what we call a degree
is a 6 × 60 = 360th of a whole arc, 1 degree is divided into 60
minutes and 1 minute into 60 seconds, and the subdivision of
the hour into minutes and seconds is just the same.

On the other hand 60 is rather a large number and not
convenient to work with. Among the numbers around 10,
12 has the greatest number of divisors:

$$12 = \begin{bmatrix} 1 \times 12 \\ 2 \times 6 \\ 3 \times 4 \end{bmatrix}$$

i.e. 1, 2, 3, 4, 6, 12, six divisors, whereas there are only 4 numbers that go into 10 without remainder: 1, 2, 5 and 10. There
are still traces of the use of the duodecimal (twelve) system:
there are 12 months in a year, 12 units in a dozen. That the
decimal system despite this got the upper hand is probably due

to the fact that man was more influenced by his games with his fingers than by utility.   The French remember that once upon a time they played with their toes.   Only such people could call 80 four times twenty (quatre-vingt); they must once have been used to the twenty number system.

Restricted as we are to the decimal system, let us see what kind of advantages this has for division.

In the first place there is a definite advantage if we want to divide by one of the divisors of 10, i.e. by 2, by 5 or by 10 itself. These will go into 10 without a remainder, as also into 2×10, i.e. into 20, into 3×10, i.e. into 30, and in the same way into all multiples of 10; also into 10×10, i.e. into 100, therefore also into 2×100, i.e. into 200, into 3×100, i.e. into 300, and so on into all the hundreds.   So we see that 2, 5, and 10 go into the tens, hundreds, thousands and so on; it is only uncertain whether they go into the ones.   For example 10 is greater than every possible value of the unit, and so whatever number there is in the one's place, that number cannot be divisable by 10 (without remainder); this is why only those numbers are divisible by 10 which have no ones.   The non-existent one is indicated by a 0, and so we get the well-known rule that only numbers ending in zero are divisible by 10.   The only unit which goes into 5 is 5 itself; that is why it is that 5 only goes into numbers ending in 0 or 5.   Finally 2 will go into 2, 4, 6, 8, and so 2 will go into those numbers which end in 0, 2, 4, 6, or 8.   These are called the even numbers.

We have exhausted all the divisors of 10, but not yet all the possibilities inherent in the decimal system.   The next unit in this system is 100.   This opens up a way to dealing with all the divisors of 100.   For example, 4 does not go into 10 but it does go into 100 since $4 \times 25 = 100$.   Therefore 4 also goes into 2×100, i.e. therefore into 200, or into any multiple of 100 without remainder, into all the hundreds, into 10×100, i.e. into 1000 and so into all the thousands and so on; it is only uncertain whether it goes into the tens and ones.   So if we want to decide whether a number, however long, is divisible by 4, we need only to examine the last two places.   For instance

$$3,478,5\underline{24}$$

is divisible by 4, because 24 is divisible by 4.   We can see this at a glance, just as though the previous five figures were not there at all. In the same way we can see at a glance that

$$312,486,\underline{434}$$

is not divisible by 4, since 4 does not go into 34 without remainder.

After the divisors of 100 we can deal with the divisors of 1000. For example 8 is not a divisor of 100, since it goes into 80 but does not go into the remaining 20 without a remainder. On the other hand it is a divisor of 1000, since 1000 can be split up like this: 800 + 160 + 40, and 8 goes into every part without remainder. For this reason 8 will go into all the thousands, tenthousands, hundredthousands, etc., without remainder; so if we want to decide whether a number, however long, is divisible by 8, we need look only at the last three places.

Now we have a recipe for deciding when a certain chosen number is a divisor of some other number: we need only to see whether this number is a divisor of 10. In this case the question can already be decided by the ones; if not, we need to go further and see if the number in question is a divisor of 100, of 1000, of 10,000, and accordingly we must examine more and more places to get an answer to the question of divisibility. Of course there are numbers which are not divisors of 10, nor of 100, nor of 1000, nor of any unit in the decimal system; as a matter of fact the majority of numbers are like that. But we can discover some regularity about these by similar investigations. The simplest case is the 9:

$$10 = 9 + 1, 100 = 99 + 1, 1000 = 999 + 1, \ldots$$

so 9 cannot be a divisor of 10, nor of 100, nor of 1000, because whichever we try to divide by it, there is always 1 left over. But just this fact that there is always 1 left over leads to a simple rule of divisibility: if we divide 10 by 9, 1 is left over, if we divide 20 by it, there will be 2 left over, if we divide 30, there will be 3 left over. In general if we divide tens by 9 as many units are left as the number of tens that we divide. In the same way, if we divide 100 by 9, 1 is left over, so if we divide 200 by it there will be 2 left over; in general in the case of the division of hundreds by 9 there are again as many units left over as the number of hundreds that we divide, and so on.

So if we want to decide whether a number is divisible by 9, it is best to split the number into ones, tens, hundreds, etc. For example

$$234 = 2 \text{ hundreds} + 3 \text{ tens} + 4 \text{ ones}$$

when we divide the 2 hundreds, there are 2 left over, when we divide the 3 tens there are 3 and when we divide the 4 ones there are 4, i.e. altogether there are

$$2 + 3 + 4 = 9$$

left over: the remainders added together give a number which is divisible by 9, therefore 234 is divisible by 9.   So here is the rule we have been looking for: a number is divisible by 9 if when we add up all its digits we get a number divisible by 9. The digits of a number which has several of them usually add up to a much smaller number than the number itself, so we can usually decide at a glance whether it is divisible by 9.   Let us examine for example the following number

$$2{,}304{,}576$$

The sum of the digits is

$$2 + 3 + 4 + 5 + 7 + 6 = 27$$

and anyone who knows his tables will know straightaway that this is divisible by 9.   On the other hand

$$2{,}304{,}577$$

is not divisible by 9 since

$$2 + 3 + 4 + 5 + 7 + 7 = 28$$

and 9 will not go into 28 without a remainder.

Our object in all we have just been doing is to avoid the difficulties caused by division.   But even in avoiding these difficulties our search has been fertile: we kept hitting on unexpected interesting relationships while doing so.   Soon we shall pluck up courage and dare to face the kind of division that cannot be carried out without remainder and this will open up new vistas towards the most daring mathematical ideas.

4. *The Sorcerer's Apprentice*

THE idea of divisibility leads on to many other interesting things, and it may be worth while to play about with these, such as, for example, the discovery that there are 'friendly numbers'. Two numbers are friendly if, when we add up the divisors of one, we get the other number as the result and vice versa. It is usual not to count among the 'proper' divisors of a number the number itself, so, for example, the proper divisors of 10 are 1, 2 and 5. Such friendly numbers are 220 and 284, because

$$220 = \begin{cases} 1 \times 220 \\ 2 \times 110 \\ 4 \times 55 \\ 5 \times 44 \\ 10 \times 22 \\ 11 \times 20 \end{cases} \text{ and } 284 = \begin{cases} 1 \times 284 \\ 2 \times 142 \\ 4 \times 71 \end{cases}$$

So the sum of the proper divisors of 220 is

$$1 + 2 + 4 + 5 + 10 + 11 + 20 + 22 + 44 + 55 + 110 = 284,$$

and the sum of the proper divisors of 284 is

$$1 + 2 + 4 + 71 + 142 = 220$$

Moreover, there are also 'perfect numbers': a number is perfect when it is equal to the sum of its own proper divisors. Such a number is, for example, 6, since its proper divisors are 1, 2 and 3, and

$$1 + 2 + 3 = 6$$

The ancients endowed such numbers with magic properties and research was started to find more perfect numbers. They did find several perfect numbers; among these we can easily check 28

$$28 = \begin{cases} 1 \times 28 \\ 2 \times 14 \\ 4 \times 7 \end{cases}$$

$$1 + 2 + 4 + 7 + 14 = 28$$

The others are much larger. These are all even numbers. They were even able to give a recipe for the construction of even perfect numbers, but we do not know to this day whether this recipe will yield any number of perfect numbers or whether it will break down somewhere. No one has yet found an odd perfect number; it is an open question whether there are any at all.

What is all this really about? Man created the natural number system for his own purposes, it is his own creation; it serves the purposes of counting and the purposes of the operations arising out of counting. But once created, he has no further power over it. The natural number series exists; it has acquired an independent existence. No more alterations can be made; it has its own laws and its own peculiar properties, properties such as man never even dreamed of when he created it. The sorcerer's apprentice stands in utter amazement before the spirits he has raised. The mathematician 'creates a new world out of nothing' and then this world gets hold of him with its mysterious, unexpected regularities. He is no longer a creator but a seeker; he seeks the secrets and relationships of the world which he has raised.

This search is so tempting just because you need practically no previous training for it except two eyes filled with curiosity. One of my little pupils of about ten came to me once with the following problem: 'I already noticed when I was in the primary school that if I add up all the numbers up to an odd number, for example up to 7, I get the same thing as if I multiply this number by its "middle". For instance the middle of 7 is 4' (this must be understood as meaning that 4 lies in the middle of the numbers, 1, 2, 3, 4, 5, 6, 7) 'and $7 \times 4 = 28$; the sum of the numbers up to 7,

$$1 + 2 + 3 + 4 + 5 + 6 + 7 \text{ is likewise } 28$$

I know this is always so, but I don't know why'. Well, I thought to myself, this is an arithmetical series all right; how should I explain it on this level? In any case I put it to the class: 'Susie has an interesting problem'. I had hardly finished speaking when the brightest little girl put her hand up and was so excited she nearly fell out of her desk. 'I'm sure it's going to be something silly, Eve; you couldn't possibly have got it in the time.' But no, she insisted that she knew. 'Well, tell us then.'

'Susie said 7 × 4, this means

$$4 + 4 + 4 + 4 + 4 + 4 + 4.$$

Susie said this instead of

$$1 + 2 + 3 + 4 + 5 + 6 + 7$$

She said 4 instead of 1, that is 3 more. But she also said 4 instead of 7, that is 3 less, and these equal out. In the same way it is true that 4 is 2 more that 2, but it is 2 less than 6 and so these too equal out. Similarly for the 4's which she said instead of the 3 and the 5 and so the two sums are really the same.'

I had to give Eve her due; I could never have explained it so well myself.

These little unprejudiced researchers make some extraordinary observations. 'It's like an exercise book,' exclaimed Mary, another little pupil. 'How do you mean?' 'Here the first and last terms equalled out, then the second and the last but one; the pages in an exercise book are joined together in the same way, the first with the last, the second with the last but one.' Pure interest was guiding these little researchers. Gauss, '*princeps mathematicorum*' is supposed to have discovered this relationship for utilitarian reasons during his primary-school days. As the story goes, Gauss's teacher once wanted a little peace and so he gave the class the lengthy task of adding up all the numbers from 1 to 100. He did not have his peace, however, since little Gauss exclaimed after a few moments: 'The result is 5,050.' The teacher had to admit that this was right, but how was it possible to calculate this so quickly? 'I noticed that $1 + 100 = 101$, $2 + 99 = 101$, $3 + 98 = 101$ and so on, —I always get 101; the last one is $50 + 51 = 101$, so after 50 such additions the terms taken from the beginning and the end finally meet in the middle. And then, of course, $50 \times 101 = 5050$.'

Little Gauss added the numbers up to an even number and so obtained a clever method for the quicker addition of such a lot of terms, just as my Susie, who reached an odd number. If we use a slightly tortuous argument we can unite the two procedures. There is a well-known joke about a person who glanced at a grazing flock of sheep and said, 'There are 357 sheep in the flock.' When they asked how he could have counted them, he replied: 'It was simple. I counted all the legs and divided by 4.' The mathematician does just that sort of thing. If, for instance, we need to add all the numbers up to a certain number, whether this is even or odd, we can calculate the double of the required sum without thinking, by adding the first to the last, the second to the last but one, etc., in the following way: let us write down the required addition twice, in two different ways. For example

$$\begin{matrix} 1+2+3+4 \\ 4+3+2+1 \end{matrix} \quad \text{or} \quad \begin{matrix} 1+2+3+4+5 \\ 5+4+3+2+1 \end{matrix}$$

In this way we write just those numbers in the same columns which we need to add up. Adding the numbers in these columns, we have

$$\begin{matrix} 5+5+5+5 \\ =4\times5=20 \end{matrix} \quad \text{and} \quad \begin{matrix} 6+6+6+6+6 \\ =5\times6=30 \end{matrix}$$

respectively, and these are the doubles of the required sums; the sums themselves we can get if we divide these by 2. Thus the results are 10 and 15 respectively, and in fact

$$1+2+3+4=10 \quad \text{and} \quad 1+2+3+4+5=15$$

We can see that in both cases we must multiply the sum of the first and the last terms by the number of terms and take half of this. In this is included Susie's as well as Gauss's result: in the case of $1+2+3+4+5+6+7$, the sum of the first and the last terms is 8, multiplying this by the number of terms $7\times8=56$ and half of this is 28. In the case of $1+2+3+\ldots+100$, the sum of the first and the last terms is 101, multiplying by the number of terms $100\times101=10100$, and half of this is 5050.

It is quite obvious (my class noticed it straightaway) that it is not only sums of consecutive numbers that can be calculated by means of this rule, but in general sums of such numbers that succeed one another by equal steps, for example (you could start with any number)

$$5+7+9+11+13$$

where every term is 2 more than the previous one, or

$$10+15+20+25+30+35$$

where the difference between any neighbouring terms is 5. For these it is also true that the sum of the first and last terms is the same as the sum of the second and the last but one and so on. Try to test them: in the first example

$$5 + 13 = 18, \quad 7 + 11 = 18$$

and for the calculation of double the first sum we need $9 + 9$ which is also 18. In the second example

$$10 + 35 = 45, \quad 15 + 30 = 45, \quad 20 + 25 = 45$$

Such series of numbers with equal intervals are called arithmetical series by mathematicians.

It is interesting that we come across the same argument in other branches of mathematics. For example, the same trick, by means of which we added the terms of an arithmetical series, is helpful in calculating areas. It is easy to calculate the area of a rectangle; this is even more simple than finding the volume of a cube: we choose a little square as our unit and we see how many of these units will make up the rectangle. Let us take for example a square inch as the unit, i.e. a small square whose length and breadth are one inch.

Let us put 8 such little squares next to each other

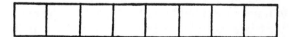

This is a rectangle. To make it less thin, let us put 3 such rows next to each other.

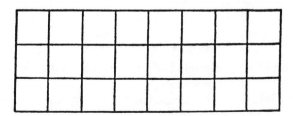

The rectangle so obtained consists of $3 \times 8 = 24$ little squares.

Conversely, if we start off with a rectangle whose length is 8″ and whose width is 3″ then there will be room in it for 3 × 8 = 24 square inches, and in general we can obtain the area of a rectangle by multiplying the lengths of two adjacent sides.

Let us note that the adjacent sides of a rectangle in addition form a right angle with each other (this is sometimes expressed by saying that they are perpendicular to each other). The right angle is an angle that you have to be very exact about, if, for example, we are building a house; neither arm of this angle leans towards or away from the other arm, as is the case with acute and obtuse angles respectively:

(for walls leaning like this would fall over very easily). The right angle preserves a proper balance.

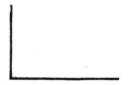

In the figure bounded by three lines, i.e. in the triangle, it is possible to have a right angle, but only one. The reader should make a few trials: however we try, the other two angles always turn out to be acute

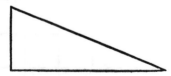

The side of a right-angled triangle which is opposite the right angle is called the hypotenuse.

Now a right-angled triangle cannot in any way be made up out of our little square units of area on account of its acute angles:

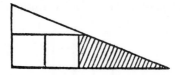

In the first row, for a start, we have left out the shaded area. The calculation of the area presents a problem.

But the problem can quite easily be solved; if we cannot calculate the area of one triangle, let us calculate the area of two. Let us fit the hypotenuse of an identical triangle upside down against the hypotenuse of our triangle. We get a rectangle

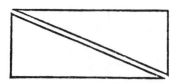

and we can calculate the area of this: we have only to multiply the two adjacent sides. The adjacent sides of the rectangle are in fact those sides of the triangle which are adjacent to the right angle. In this way we can calculate double the area of the triangle: we get the area of one triangle by dividing the result by 2. So we calculate the area of a right-angled triangle by multiplying the lengths of the sides adjacent to the right angle and dividing by 2.

It becomes quite obvious that the main argument here is the same as in the summing of the terms of an arithmetical series if we follow the exposition of the mathematician Euclid, who bequeathed to the world a marvellously complete mathematical work 2000 years ago. Euclid dresses up the properties of numbers in geometrical clothes: with him the symbols for

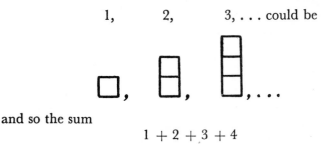

1,   2,   3, . . . could be

and so the sum

$$1 + 2 + 3 + 4$$

could be represented by means of a 'triangle with steps' like this

The trick in which we put the sum written backwards underneath now becomes that of fitting another triangle with steps on
top of the first one, like this

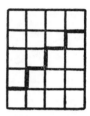

In this way 1 comes on top of 4, 2 on top of 3, 3 on top of 2 and
4 on top of 1.   The squares that are on top of each other everywhere amount to 5, altogether giving $4 \times 5 = 20$, corresponding to the fact that the rectangle so formed has a length 4 units
and width 5 units, i.e. the rectangle takes up 4×5 units of area.
This is the double of the sum in question; the sum itself is the
half of this, since the area of each triangle with steps is half the
area of the rectangle.   It should now be quite clear that we
have gone through the same argument, once in arithmetical,
once in geometrical language.   We shall see that this argument
has still a great many more variations.

5. *Variations on a fundamental theme*

UNDER what circumstances do we have to sum numbers from 1 onwards? The following, seemingly quite different problem, leads also to this process.

We have already come across triangles and quadrilaterals; in general, figures enclosed by straight lines are called polygons.

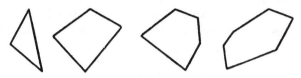

All these drawn above are so-called 'convex' polygons. They are not indented anywhere as are those below

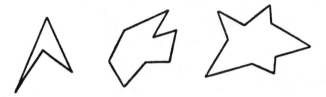

The latter differ from the former in that you can produce a side of the latter and thus slice them in two:

The reader should try to convince himself that this cannot be done with the former, for we must be clear about this difference. In what follows we shall be dealing only with convex figures. (We shall make the same distinction among solids too.)

The lines joining non-neighbouring vertices are called diagonals (since the line joining two neighbouring vertices is not

a diagonal, but a side.)   For example I shall draw a few
diagonals in the polygon below

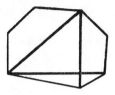

Now the problem is this: given a polygon, say an octagon,
how many diagonals can I draw in it?   Even if I draw them all
in, it is not so easy to count them, for they cover the figure so
thickly.

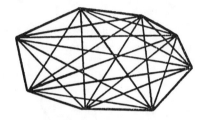

The problem is simplified if we do not distinguish between
neighbouring and non-neighbouring vertices, and so count in
the sides for the time being.   We know in any case that there
are 8 sides, so we shall have to subtract 8 from the result.

In this form the problem can be put in the following way.
Given the 8 vertices of an octagon,

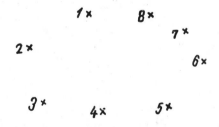

in how many ways can we join these up in pairs?   Two ways
appear to be indicated for the solution.   One of them is that we
join point 1 with the other seven, in this way we get 7 joins.

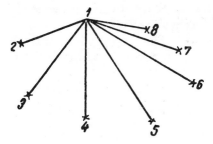

Then we join point 2 with the others, except for 1, with which it is already joined. In this way we add another 6 joins to those we have already.

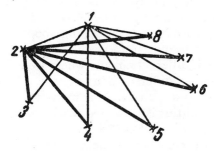

Now we join point 3 with the others except for the two points already dealt with; in this way we get 5 new joins; similarly by joining point 4 we get 4 new joins, joining point 5 will yield 3 more joins, point 6, 2 more and point 7 just one, while point 8 has already been joined to all the other points and so it will not yield any new joins. Altogether therefore we get

$$7 + 6 + 5 + 4 + 3 + 2 + 1$$

joins, or writing it the other way round

$$1 + 2 + 3 + 4 + 5 + 6 + 7$$

lines.

The other way of counting these joins is to see how many you can draw from each vertex, independently of all the others. Of course you can draw 7, since each vertex can be joined to all of the 7 other vertices. Now we can argue further in the following way: if from one vertex we can draw 7 lines, from 8 vertices we can draw 8 × 7 such lines. This is wrong though, since every line joins 2 vertices; so for example the line joining the vertices 1 and 6 was included when we were counting the lines drawn from the vertex 1 as well as when we were counting those from the vertex 6.   The mistake is simply that we counted every join twice.   The correct result is the half of $8 \times 7 = 56$, i.e. 28.

We must reach the same result with both ways, so

$$1 + 2 + 3 + 4 + 5 + 6 + 7$$

is therefore the half of 8×7; this is again the result obtained by my pupil Susie.

But the theme can be varied even further.   The problem that has arisen can be formulated differently in the following way: since any line joins two vertices, the question really is in how many ways can 2 vertices be chosen out of 8?   So we see that the fact that we have been talking about vertices is really quite irrelevant; we could just as well pose the problem of a bag with 8 balls all of different colours where we ask in how many ways we can choose different pairs of balls.   Or if we wish to divide 8 children into pairs, in how many ways can we choose the first pair?   All this is expressed mathematically by saying: how many combinations of 2 can be generated out of 8 elements?

If we denote the elements by the numbers 1, 2, 3, 4, 5, 6, 7, 8, then the following are the combinations of 2 (or, more simply, pairs) that can be generated from them

1 2	2 3	3 4	4 5	5 6	6 7	7 8
1 3	2 4	3 5	4 6	5 7	6 8	
1 4	2 5	3 6	4 7	5 8		
1 5	2 6	3 7	4 8			
1 6	2 7	3 8				
1 7	2 8					
1 8						

We can see very well that the number of these pairs is (from right to left)

$$1 + 2 + 3 + 4 + 5 + 6 + 7$$

On the other hand we might have argued that any element can be paired with the remaining 7, so the 8 elements would yield 8×7 pairs; but we have counted every pair twice, first when we were pairing the first term and then when we were pairing the second.   The correct result is therefore again half of 8 × 7. All these different starting points lead to the same end-result. I cannot help expressing this in the form of a formula. The only thing I must warn the reader about is that in mathematics the bracket does not indicate something of lesser importance; we should note that mathematicians put in brackets things whose coherence they want to emphasize. For example (2 + 3) × 6 means that we must multiply by 6 the result of the addition 2 + 3, i.e. 5, whereas if we wrote it without brackets 2 + 3 × 6 would mean that we must add to 2 the result of the multiplication 3 × 6 [there is a convention that multiplication 'ties more closely' than addition, and so the latter need not be written as 2 + (3 × 6)]. Everybody knows that the half of 4, of 6, of 10 can be written as $\frac{4}{2}, \frac{6}{2}, \frac{10}{2}$ respectively and that, in general, division can be expressed in this 'fractional' form. Then, if we denote by n the number up to which we have been summing all the numbers, the sum of the first and the last terms will be $1 + n$. So we must multiply this by the number of terms, i.e. by n and then divide this by 2. In other words all the variations of our fundamental theme can be condensed in the following formula

$$1 + 2 + 3 + \ldots + n = \frac{(1 + n)n}{2}$$

Mathematics is really a language, a queer language which speaks entirely in symbols. The above formula is only a symbol, and means nothing by itself; everybody can substitute into it his own experiences. For one it might mean the counting of the diagonals of a polygon, for another the counting of the number of possibilities for choosing the leading pair among his pupils. The writing down of a formula is an expression of our joy that we can answer all these questions by means of one argument.

Postscript on geometry without measurements

We have now become aware of two new themes, one geometrical and one arithmetical. I should like to follow up the geometrical one a little further at first.

Let us have another look at the figure showing the octagon with all its diagonals. You cannot make head or tail of this figure because the diagonals cut across each other all over the place, and there is an enormous number of intersections; it is a good job that the polygon is convex, for its vertices are all on the outside and so we cannot mix them up with the intersections. The whole thing would be easier to see if the diagonals were made of elastic string fixed at the vertices and then you could pull the pieces of string out into space.   One person could hold each one of these diagonals, the second one would be pulled a little higher than the first, the third a little higher than the second and so on.   In this way they would not intersect each other and they could be counted, since this stretching does not alter their number.

There is a special branch of Geometry, called Topology, which deals with those properties of figures which are not altered if we make the figures of elastic and stretch them or compress them in any way whatsoever.   It is a strange thing that this study is classed as part of geo*metry*, since there is no question of *metric* properties, i.e. of measurements, because the distances and angles are altered during the stretching.   From our point of view what makes these considerations interesting is that they are new and we know their origin: before our very eyes we picture a whole branch of mathematics being born out of a game.

This game was a puzzle in connection with the bridges at Königsberg.   The River Pregel, which runs through Königsberg, has two islands.   Those two islands are joined to each other and to the shores by means of seven bridges, as shown on the diagram below

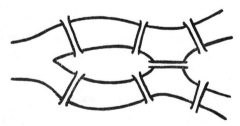

The puzzle is whether it is possible, starting at any point, to work out a walk so that we get back to our starting point having walked across all the bridges, but only once over each bridge.   The reader should try this for himself a few times; in

the meantime it should become clear to him that the problem would not be altered at all if the bridges leading to the same island or shore converged at the same points (this would merely cut out the walks on the banks of the river), so that the map would look like this

It would of course be quite silly to have two bridges connecting the same points of the shore and the island, but we might suppose that one was built for pedestrians and the other for cars. This consideration enables us to sketch a simpler schematic drawing

and the problem can now be re-formulated by asking whether this figure can be drawn with one stroke of a pencil without lifting the pencil off the paper (since our walker cannot rise into the air) in such a way that no part of the figure should be drawn twice and that at the end we should get back to the starting point. This puzzle probably sounds familiar; it is usual to pose the problem in connection with envelopes like the following

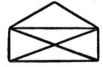 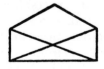

It is obvious that these problems belong to Topology, for whether such a figure can or cannot be drawn by one pencil stroke is not influenced by imagining the whole figure made of pieces ot elastic which can be stretched, compressed and generally

deformed; though we must not tear it or stick any parts of it together.

The great Euler gave a simple answer to all questions of this kind. If a figure can be drawn by one pencil stroke leading back to its starting point, then the pencil must start from the starting point, and must also return to it, and every time it comes to a vertex, it must also leave that vertex in order to go farther. So every line coming to a vertex has a mate, namely a line leaving the vertex, and therefore at every vertex there must be an even number of lines meeting. It can be proved that this statement is sufficient; a figure can always be drawn by means of one line ending at its starting point if there is an even number of lines meeting at every vertex.

According to this the problem of the walk in Königsberg is insoluble; every vertex in the corresponding diagram makes this impossible, since on the far-left vertex there are 5 lines meeting and at the remaining three vertices there are 3 lines meeting at each and these are all odd numbers.

On the other hand the first envelope can be drawn if we do not insist on our pencil coming back to the original starting point, since at the upper vertices there are 4 lines meeting, in the middle also 4 and these are even numbers; only the two lower vertices could spoil things, since here there are only 3 lines meeting. But if we allow the line to start at one of these vertices and end at the other, then we can draw it in the following stages.

The second envelope is a hopeless case, since it has more than two recalcitrant vertices: it is only at the uppermost and at the central vertex that an even number of lines meet; at the remaining vertices three lines meet at each.

This is the game from which Topology originally started, but it must not be thought that it has remained at the playful stage; it has become a very serious branch of Science, so that other Sciences make use of it to a considerable extent. For example Physics uses Topology in the description of circuits, Organic Chemistry uses it in connexion with molecular models; in general, topological considerations occur in every case where we want to produce a structure irrespective of magnitudes.

It is worthwhile to ponder for a while on what kind of geometrical notions go by the board in Topology. Such notions are those of congruence and of similarity. Congruence of triangles plays the most important part in Geometry, since other plane figures can be split up into triangles; polygons can be split by means of diagonals

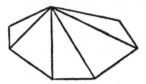

Even a circle can be considered as made up of triangles approximately (we shall come back to this approximation), if we draw the radii quite close to one another

so that each bit of arc looks almost like a straight line. (I know that this 'almost' is an unpleasant school memory, a certain sense of uncertainty is associated with it. I promise that later I shall give a precise meaning to it.)

Two triangles are congruent if you can put one on top of the other so that one exactly covers the other; for example the two triangles below are like that.

The reader can convince himself of this by cutting them out of paper and turning them into identical positions. When one is placed on top of the other, all six elements (three sides and three angles) are covered exactly. To achieve this it is enough that certain pairs of elements should be equal, for example, when two sides and the included angle of one triangle are equal to the two sides and included angle of the other triangle.

Since we know that the heavily drawn elements are equal, we can place the equal angles on top of each other, and then the end-points of the adjacent sides will also lie on top of each other; the third side of the uppermost triangle lies between these two points, and so it cannot do anything else but lie on the third side of the lower triangle, with the angles likewise covering each other.

Two triangles are similar if their shape is similar but they may be different sizes; one might be a small edition of the other.

We can imagine this by thinking that we have taken a picture of the large triangle and the camera has made the triangle smaller.   It will be readily seen that we can choose the length of one of the sides of the little triangle at will for we are supposing that our camera is capable of making things as small as we like.   We must certainly also assume that it does not distort; namely that it makes the other two sides smaller in just the same ratio.   On the small figure the sides lean away from each other to the same extent, so that the angles are not altered at all.   We see therefore that the sides of similar triangles are bigger or smaller than each other in just the same ratio (we express this by saying that they are proportional) and their corresponding angles are equal.

For this, on the other hand, it is enough that two of the angles should be equal.   Since, if we wish to draw a triangle similar to a given triangle,

we can take any side of the triangle, for example the lower one, as smaller or larger

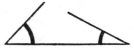

After this we must draw the two lower angles of the triangle

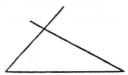

and now we have used everything up: if we produce the arms of the two angles, the triangle will be closed

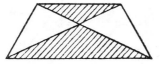

and so this must be the similar triangle required.   Thus similarity is really decided by the equality of two sets of corresponding angles.

In Geometry we keep coming across congruent and similar figures; for example in the isosceles trapezium, below, the white triangles are congruent and the shaded ones are similar:

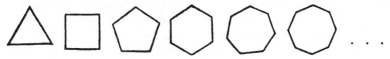

Now in Topology we cannot deal with congruence or similarity; if we stretch and compress figures, both the size and the shape are altered, straight lines might be bent into curves, they might even escape out of their plane.

It is interesting, however, that topological considerations can be used to decide the question of how many regular solids there are, although the 'regularity' of solids is intimately connected with congruence, i.e. with measurements.   A convex solid is regular if it is bounded by plane figures with equal sides and equal angles like these:

△ □ ⬠ ⬡ ⬠ ○ . . .

All these bounding figures are congruent, and at every vertex the same number of them meet. The way Topology deals with this problem is by confining itself to certain properties of regular solids, namely that every face is bounded by the same number of edges, and that at every vertex the same number of edges meet. These properties have nothing to do with size or shape. In this way it is possible to prove by means of topological tools alone that there can be only five solids satisfying even these few requirements. To prove that there really are five bodies satisfying these conditions we use the branch of Geometry that can measure. Three of these bodies are bounded by triangles, the well-known cube is bounded by squares, and one is bounded by pentagons.

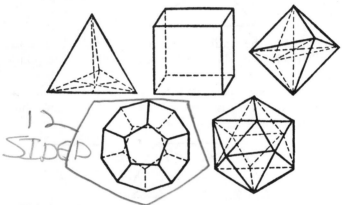

This is quite a surprising discovery since in the plane there is no reason why there should not be regular polygons with as many sides as we like. The series shown on p. 39 could be continued indefinitely. What we imagine in the plane therefore must not be transferred without thinking to three dimensions. In space a lot of things happen differently.

It is again worth while to ponder over these things a little. We did anticipate that we should meet new phenomena in space, since it is so much easier to move about in space than if we are restricted to moving in a plane. We expected that there would be more possibilities than in a plane, for example a greater variety of kinds of regular solids. But this does mean that the larger number of possibilities create conditions harder to fulfil in some cases, since there are more possibilities in the determination of the conditions themselves. At one vertex of a three-dimensional solid not only two edges can meet, as in the case of plane figures, but any number of edges, and what is more, any number of faces; you could have 30 edges meeting at a vertex, 3 at another, one face could be a triangle, while another could be a polygon with 30 sides. It is an extremely severe restriction on a solid not to be able to make use of all these possibilities, confining it to an equal number of edges at every vertex and round every face, and allowing it only a single choice. Altogether only five solids can cope with these restrictions.

It is strange that I thought of Topology in connexion with carrying out the addition $1 + 2 + 3 + \ldots + n$ in a more sensible way. This shows too that Mathematics is an organic whole: wherever we touch it, connecting links from all other branches come crowding into our minds.

6. *We go through all possibilities*

THE teacher is not likely to trouble very much about how many possible ways there are of choosing pairs of children out of his class; he will solve the problem of pairing quite satisfactorily by considering friendships and hostilities among the children. But the young research worker who is still full of fresh curiosity will want to go through all possible ways. In one of my first forms in the secondary school when we were discussing a point about multiplying by 357 namely, that we could start the multiplication either with the units or with the hundreds, somebody immediately asked whether you could not start with the tens. When I replied that you could, but that you would have to be extra careful about where you wrote down your partial products, they straightaway wanted to know in how many ways you could actually carry out a certain given multiplication. On account of this I was obliged to make a little excursion into the theory of combinations, this being the branch of Mathematics which deals with the number of possible arrangements.

There is hardly a child who would not be interested to know how many different flags you could make with three different colours. With one colour you can of course make only one flag:

and we can add a coloured strip to this in only two ways (if we want to use every colour once only): we have to put it either above or below the first strip:

How is it possible to add another colour? You could put it

above, in between or below.   From the two-colour flag on the left we can then make three new flags:

and we can do likewise with the flag on the right:

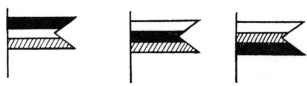

So out of three colours we can make altogether $2 \times 3 = 6$ flags. From this we can proceed to four colours in just the same way. The fourth colour can be put above the first colour, between the first and the second, between the second and the third, finally under the third colour, and this can be done with any of the three-colour flags.   In this way out of every three-colour flag we can make four four-colour flags, for example out of the first we can make:

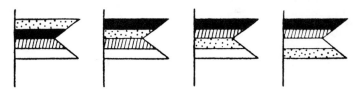

In this way, therefore, out of $2 \times 3 = 6$ three-colour flags, we can make altogether $2 \times 3 \times 4 = 6 \times 4 = 24$ flags.   We can put the 1 in as a factor, since it does not make any difference, and then we notice the following beautiful regularity:

The number of one-colour flags   .   .   1
The number of two-colour flags   .   . $1 \times 2 = 2$
The number of three-colour flags .   . $1 \times 2 \times 3 = 6$
The number of four-colour flags   .   . $1 \times 2 \times 3 \times 4 = 24$

It is quite clear that the procedure will be the same even if we are not dealing with colours.   For example you can serve out soup for five children in $1 \times 2 \times 3 \times 4 \times 5 = 120$ different orders, or any six 'elements' can be arranged, or 'permuted', in $1 \times 2 \times 3 \times 4 \times 5 \times 6 = 720$ different ways. In the expression

$$1 \times 2 \times 3 \times 4 \times 5 \times 6$$

the following command is expressed: multiply all the numbers from 1 to 6 but no further! This is usually abbreviated by writing down only the last factor and putting an exclamation mark after it, so that a short way of writing the above would be

$$6!$$

Since we are dealing with factors, we say 'six factorial' when we read this sign. For example:

$$1! = 1, \qquad 2! = 1 \times 2, \qquad 3! = 1 \times 2 \times 3 \qquad \text{and so on.}$$

The value of the factorial depends of course on how far we go on multiplying, so we have again come across a function. Let us quickly construct its 'temperature chart', representing on a horizontal line that number at which we stop the multiplications, and upwards the corresponding factorial

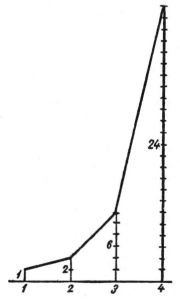

On the opposite page are shown the powers of 2.

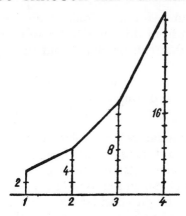

We can see that at first the curve of the factorials remains below the curve of the powers (please look at the portion between 1 and 2, for example), but afterwards it rises above the power curve and shoots up much more steeply.   And this is not only true of the powers of 2; the factorial curve rises more steeply than the curve of any power whatsoever.   Of course this is quite natural; whatever the value of the base, for example 100, when we raise it to a power, we always multiply by this same 100.   The first 99 factors of the factorial are of course smaller than 100, but after a 100 they get bigger, and we multiply by 100, 101, 102, 103, . . . and so sooner or later they will get the upper hand.

We were able to work out the number of two-, three- and four-colour flags gradually from the number of the one-colour flag by means of the beautifully regular series

$$1 \times 2, \qquad 1 \times 2 \times 3, \qquad 1 \times 2 \times 3 \times 4, \ldots$$

Other problems in the theory of combinations also lead to similarly beautiful results.   For example we already know how to pick out all the possible pairs from a certain number of elements: we have shown that out of 8 elements you can do this in

$$\frac{8 \times 7}{2}$$

different ways; out of 15 elements in

$$\frac{15 \times 14}{2}$$

different ways, and so on. Would it not be possible step by step to construct from this the number of possible groups of 3, of 4, and of 5 that can be chosen out of a given number of elements?

Let us see in how many ways we can join a third element to one of the pairs, for example to

$$1, 2,$$

from the pairs that can be constructed out of the elements 1, 2, 3, 4, 5, 6, 7, 8. In this case we are going to disregard the order of choosing the elements; we shall be concerned only with whether a certain element has found its way into a group or not (for example one might consider the problem of appointing a committee of 3 out of a group of 8 people, the only point being who shall be appointed). Therefore to the pair 1, 2 we could add any one of the remaining 6 elements and so we obtain the following six groups of three:

$$
\begin{array}{c c c}
1 & 2 & 3 \\
1 & 2 & 4 \\
1 & 2 & 5 \\
\hline
1 & 2 & 6 \\
1 & 2 & 7 \\
1 & 2 & 8 \\
\end{array}
$$

(please ignore the line in the middle for the time being).

In the same way we can enlarge every other pair into a group of three in six different ways, for example the pair 2, 5 can be enlarged into the groups

2 5 1		1 2 5
2 5 3		2 3 5
2 5 4	or arranged in order of magnitude	2 4 5
2 5 6		2 5 6
2 5 7		2 5 7
2 5 8		2 5 8

At first it might appear that we can construct six times as many groups of three out of 8 elements as we can construct pairs. But among these there will be identical groups, for example 1, 2, 5 can be constructed out of 1, 2 as well as out of 2, 5 (I have underlined them in both places), and what is more it must also occur among the enlargements of the pair 1, 5, since we can

add 2 to this pair as the third element.   It is obvious that every
group of three will be constructed three times, i.e out of each
pair which we obtain by leaving out an element from that group
of three.   For example if we leave out an element from 2, 3, 5,
we are left with one of the following pairs:

$$2 \quad 3$$
$$2 \quad 5$$
$$3 \quad 5$$

and the group 2, 3, 5 is constructed out of the first pair by joining a 5, out of the second by joining a 3, out of the third by
oining a 2.   If, therefore, we wish to obtain every group only
once then we must divide by 3.   So finally we see that to obtain
all groups of 3 that can be chosen out of 8 elements, we must
multiply the number of pairs that can be chosen out of 8 elements by 6, then divide the result by 3.   We already know that
the number of pairs is $(8 \times 7)/2$; we can multiply this by 6 in
such a way that the division by 2 is left to the last

$$\frac{8 \times 7 \times 6}{2}$$

We still need to divide by 3.   To divide by 2 and then to divide
by 3 is the same as to divide by 2×3 (for example $12/2 = 6$
and $6/3 = 2$, and if we divide 12 by $2 \times 3 = 6$, we get 2).
So finally—putting in an inessential factor 1 in the denominator
for aesthetic reasons—from 8 elements we can choose

$$\frac{8 \times 7 \times 6}{1 \times 2 \times 3}$$

groups of three.   In the same way we can see that out of 12
elements we can choose

$$\frac{12 \times 11 \times 10}{1 \times 2 \times 3}$$

groups of three and out of 100 elements

$$\frac{100 \times 99 \times 98}{1 \times 2 \times 3}$$

groups of three.

Once we know the number of groups of three, we can go on
to groups of four in just the same way.   Let us consider again
8 elements, then out of every group of three we can construct 5

groups of four by adding one of the remaining elements: for example out of the group

$$1 \ 2 \ 3$$

we can construct the groups

$$1 \ 2 \ 3 \ 4$$
$$1 \ 2 \ 3 \ 5$$
$$1 \ 2 \ 3 \ 6$$
$$1 \ 2 \ 3 \ 7$$
$$1 \ 2 \ 3 \ 8$$

According to this we should get five times as many groups of four as there were groups of three, obtaining every group four times. For example

$$1 \ 2 \ 3 \ 4$$

can be obtained from

$$1 \ 2 \ 3 \quad \text{by joining} \quad 4$$
$$\text{from} \quad 1 \ 2 \ 4 \quad \text{by joining} \quad 3$$
$$\text{from} \quad 1 \ 3 \ 4 \quad \text{by joining} \quad 2$$
$$\text{and from} \quad 2 \ 3 \ 4 \quad \text{by joining} \quad 1$$

so we should divide the result by 4. The number of groups of three was

$$\frac{8 \times 7 \times 6}{1 \times 2 \times 3}$$

so we must multiply this by 5 and divide it by 4. The number of groups of 4 will be

$$\frac{8 \times 7 \times 6 \times 5}{1 \times 2 \times 3 \times 4}$$

The reader will surely see the rule emerging. The number of groups of 7 that can be chosen out of 10 elements is

$$\frac{10 \times 9 \times 8 \times 7 \times 6 \times 5 \times 4}{1 \times 2 \times 3 \times 4 \times 5 \times 6 \times 7}$$

This too is a beautifully regular result: when we choose groups of 7 we have seven factors in the numerator as well as in the denominator, only in the denominator the factors proceed upwards from 1, while in the numerator they proceed downwards from 10, if we are choosing out of 10 elements.

For example, the number of single elements that can be

chosen out of 5 elements is 5/1 = 5, which is obvious; the number of groups of three that can be chosen out of three elements is

$$\frac{3 \times 2 \times 1}{1 \times 2 \times 3} = \frac{6}{6} = 1$$

and this is also quite obvious, since out of three balls you can choose all three in one way only. There is also only one way of withdrawing our hand without choosing any balls at all, no matter how many balls there are in the bag. So let us agree that, however many elements we are choosing from, the number of zero combinations will be 1. Therefore the number of combinations can be expressed in the following table:

	none	one	two	three	four
		choose			
from 1	1	$\frac{1}{1}=1$	—	—	—
from 2	1	$\frac{2}{1}=2$	$\frac{2 \times 1}{1 \times 2}=1$	—	—
from 3	1	$\frac{3}{1}=3$	$\frac{3 \times 2}{1 \times 2}=3$	$\frac{3 \times 2 \times 1}{1 \times 2 \times 3}=1$	—
from 4	1	$\frac{4}{1}=4$	$\frac{4 \times 3}{1 \times 2}=6$	$\frac{4 \times 3 \times 2}{1 \times 2 \times 3}=4$	$\frac{4 \times 3 \times 2 \times 1}{1 \times 2 \times 3 \times 4}=1$

and so on.

We can arrange these results in the following order, if we add another 1 at the highest position, corresponding to the fact that there is only one way of withdrawing our hand with nothing in it from an empty bag, i.e. the number of zero combinations out of a zero number of elements can also be considered as 1:

```
                1
             1     1
          1     2     1
       1     3     3     1
    1     4     6     4     1
```

This triangular figure is called the Pascal Triangle. It has a number of interesting properties. It is natural that it is symmetrical, i.e. that its left-hand side is a mirror image of its right-hand side, since for example out of 3 balls you can pull out 1 in the same number of ways as leaving 2 in the bag. Similarly, if we are constructing pairs out of 5 elements, every time we construct a pair, we also construct a group of three out of the remaining elements, i.e. in the case of 5 elements the number of pairs is the same as the number of groups of three. And it is just these numbers that we can see in the Pascal Triangle as mirror images of each other.

Another property yields a simple rule for constructing the other rows in the Pascal Triangle. I have not written 2 in between the 1 and the 1 without reason; it is because $1 + 1 = 2$. In the same way 3 is between the 1 and the 2 and $1 + 2 = 3$ and so on. This goes on like that quite regularly, and so, since $1 + 4 = 5$ and $4 + 6 = 10$, the row following the last one in the figure is

$$1 \quad 5 \quad 10 \quad 10 \quad 5 \quad 1$$

and similarly the one after that will be

$$1 \quad 6 \quad 15 \quad 20 \quad 15 \quad 6 \quad 1 \quad \text{and so on.}$$

The proof is also quite simple, but let a trial verification suffice. The first 15 is in the place of the number of pairs that can be constructed out of 6 elements. The number in question is

$$\frac{6 \times 5}{1 \times 2} = \frac{30}{2}$$

and this is really 15.

It follows from this that the sum of the terms in each row is double the sum of the terms in the previous row. For example let us construct the row after the last one we wrote down. It is done in the following way:

$$1 \quad \underbrace{1 + 6} \quad \underbrace{6 + 15} \quad \underbrace{15 + 20} \quad \underbrace{20 + 15} \quad \underbrace{15 + 6} \quad \underbrace{6 + 1} \quad 1$$

and we can see clearly that in this row every term of the row

$$1 \quad 6 \quad 15 \quad 20 \quad 15 \quad 6 \quad 1$$

occurs exactly twice.

This throws further light on another property of the Pascal Triangle: adding the terms of a row we obtain the successive powers of 2. Since this is the case in the beginning (apart from the uppermost 1): i.e. $1 + 1 = 2 = 2^1$, $1 + 2 + 1 = 4 = 2^2$, we need not trouble to look any farther; if this property is true for one row, then it will be 'inherited' by the next row. We know that the sum of the terms of each row is twice as much as the sum of the terms in the previous row, and if we multiply any power of 2 by 2, we shall get a product $2 \times 2 \times 2 \times \ldots 2 \times 2$ with one more 2 in it, i.e. we shall get the next power of 2.

This kind of proof, which is based entirely on the construction of the natural number series, is called mathematical induction. The natural number series begins with 1, and by continuing to count one more, we can reach any member of the series. The idea of mathematical induction is simply that if something is true at the beginning of the number series, and if this is 'inherited' as we proceed from one number to the next, then it is also true for *all* natural numbers. This has given us a method to prove something for *all* natural numbers, whereas to try out all such numbers is impossible with our finite brains. We need prove only two things, both conceivable by means of our finite brains: that the statement in question is true for 1, and that it is the kind that is 'inherited'.

This is a most important lesson, namely that the infinite in mathematics is conceivable by means of finite tools.

Those who like to play about with multiplications will be familiar with the first few rows in the Pascal Triangle. If we construct the powers of 11 successively, we find that

$$11^1 \qquad\qquad = \qquad 1 \quad 1$$

$$11^2 = 11 \times 11$$
$$\underline{11}$$
$$121 \qquad\qquad = \qquad 1 \quad 2 \quad 1$$

$$11^3 = 121 \times 11$$
$$\underline{121}$$
$$1331 \qquad\qquad = \qquad 1 \quad 3 \quad 3 \quad 1$$

$$11^4 = 1331 \times 11$$
$$\underline{1331}$$
$$14641 \qquad\qquad = 1 \quad 4 \quad 6 \quad 4 \quad 1$$

The figures in the results are the very numbers in the Pascal Triangle. Those who had a good look at the multiplications will know straight away why this is so; when we added the partial products, we carried out just the same additions as in the construction of the rows in the Pascal Triangle (in the case of 11^5 this is spoilt by the fact that in the addition of the partial products there is carrying to be done

$$11^5 = 14641 \times 11$$
$$14641$$
$$\overline{161051}$$

whereas the corresponding row in the Pascal Triangle is
1   5   10   10   5   1)

11 is really $10 + 1$
$121 = 100 + 20 + 1 = \mathbf{1} \times 10^2 + \mathbf{2} \times 10 + \mathbf{1}$
$1331 = 1000 + 300 + 30 + 1 =$
$$\mathbf{1} \times 10^3 + \mathbf{3} \times 10^2 + \mathbf{3} \times 10 + \mathbf{1}$$

and so on.   So the numbers in the Pascal Triangle occur as coefficients of descending powers of 10 in the expression of the powers of $10 + 1$.   The second term in $10 + 1$ is 1, and every power of 1 is still 1 (since $1 \times 1 = 1$), and so it does not appear that the powers of the second term come into the expression at all.   But we can smuggle them in in the following way:

$11^3 = 1331 = 1000 + 300 + 30 + 1 =$
$$\mathbf{1} \times 10^3 + \mathbf{3} \times 10^2 \times 1 + \mathbf{3} \times 10 \times 1^2 + \mathbf{1} \times 1^3$$

We see that while the powers of the first term decrease, the powers of the second term increase.   The importance of this lies in the fact that we can generalize this expression to yield the expansion of powers of other sums of two terms.   For example

$$7^3 = (5+2)^3 = \mathbf{1} \times 5^3 + \mathbf{3} \times 5^2 \times 2 + \mathbf{3} \times 5 \times 2^2 + \mathbf{1} \times 2^3$$

After what we have seen, it would not be difficult to prove this in the general case, but let us be satisfied with a numerical check:

$$
\begin{array}{lll}
1 \times 5^3 = 5 \times 5 \times 5 = 25 \times 5 & = & 125 \\
3 \times 5^2 \times 2 = 3 \times 5 \times 5 \times 2 = 15 \times 10 & = & 150 \\
3 \times 5 \times 2^2 = 3 \times 5 \times 2 \times 2 = 3 \times 10 \times 2 = & & 60 \\
1 \times 2^3 = 2 \times 2 \times 2 = 4 \times 2 & = & 8 \\
\hline
& & 343
\end{array}
$$

and in fact $7^3 = 7 \times 7 \times 7 = 49 \times 7 = 343$

This discovery is again a very convenient one; quite often it is easier, instead of raising a single number to a power, to split this number into two terms whose powers are easy to calculate. For example some people do not like multiplying by 7; when we calculate the expanded form of $(5 + 2)^3$, we need only to do easy multiplications, namely those by 5 and by 2 (these, if possible, should be carried out in such a way as to yield as many multiplications by 10 as possible; multiplication by 10 is really child's play).

Another word for 'two terms' is binomial, this is why this expansion is called the Binomial Theorem, and the numbers in the Pascal Triangle are called the binomial coefficients.

It is the second power which is most often needed. The second row in the Pascal Triangle is

$$1 \qquad 2 \qquad 1$$

According to this, if we wish to calculate $(5 + 3)^2$, then these are the numbers by which we have to multiply; powers of 5 decreasing from 2, and powers of 3 increasing up to 2, will enter into the expansion, so that

$$(5 + 3)^2 = 1 \times 5^2 + 2 \times 5 \times 3 + 1 \times 3^2$$

or leaving out the superfluous factors 1:

$$(5 + 3)^2 = 5^2 + 2 \times 5 \times 3 + 3^2$$

Thus we reach the well-known rule (of which the reader may have unpleasant memories!): to raise the sum of two terms to the second power, we add the second power of the first term to double the product of the first two terms, and then to the second power of the second term.

Of course we could have seen this in a much simpler way, for example by looking at it geometrically. We know that the area of a rectangle is the product of the lengths of the two adjacent sides. So, conversely, if we have a product, we can represent it by means of the area of a rectangle, the lengths of the adjacent sides of which are the factors. For example here is the representation of the product 3×5

and that of the product $5^2 = 5 \times 5$

This is of course a square.   That is why the second power is also called a square.

Let us now represent the expression $(5 + 3)^2$

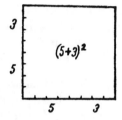

Here the separate identity of the terms has all but been lost sight of, but dividing up the square in the manner shown will throw more light on the role played by the terms.

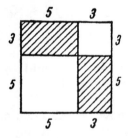

Out of the pieces thus produced we see that the area of the larger square is 5^2, the area of the smaller square is 3^2.   Apart from these there are two more rectangles with areas 5×3 units each, making up the original large square.   Therefore in fact

$$(5 + 3)^2 = 5^2 + 2 \times 5 \times 3 + 3^2$$

This is just as clear as the figures in Hindu textbooks.   The Hindus do not believe in a lot of verbiage.   They state the theorem: a sum of two terms can be squared in such and such a way. Then they write: 'See below', and they draw a figure which explains it all:

$a \times b$	b^2
a^2	$a \times b$

Those who have eyes to see, will see.

7. Colouring the grey number series

THE Hindus have been excellent mathematicians from time immemorial, and they have abilities quite peculiar to themselves in this field. I heard the following anecdote about one of their scientists: One of his European friends once asked him jokingly whether the registration number, 1729, of the taxi he had just been in was an unlucky number; he replied quite naturally: 'Oh no, on the contrary, this number 1729 is a very interesting number. It is the first number that can be expressed in two ways as a sum of two cubes, since $10^3 + 9^3$ as well as $12^3 + 1^3$ are both 1729.'

To the Hindus even the four-digit numbers are somehow like personal friends endowed with their own special peculiarities. In our primary schools the small numbers are treated in this individual sort of way: for the young pupil 2 is not just one of the many grey numbers, but has an individuality which he has learnt to recognize in many of its aspects: it is the first even number, it is $1 + 1$, it is half of 4 and so on. But whether we colour our numbers up to 10 or up to such large numbers as the Hindus, all this is a very tiny fraction of the infinite number series, which goes rolling on and on in a grey sort of way.

We know that there are even numbers; yes, every other number is even:

$$1, \underline{2}, 3, \underline{4}, 5, \underline{6}, 7, \underline{8}, 9, \underline{10}, 11, \underline{12}, \ldots$$

In the same way every third number is divisible by 3,

$$1, 2, \underline{3}, 4, 5, \underline{6}, 7, 8, \underline{9}, 10, 11, \underline{12}, \ldots$$

every fourth number by 4,

$$1, 2, 3, \underline{4}, 5, 6, 7, \underline{8}, 9, 10, 11, \underline{12}, \ldots$$

and so on. These are only waves, some small, some larger, but once they have started, they go rolling on with monotonous regularity. Is there nothing unexpected here, no individual unpredictability, which would brighten up this monotony?

Fortunately there is: it is the distribution of prime numbers, which is quite unpredictable and quite impossible to squeeze

56

into any regular pattern. Let us remind ourselves what divisibility is:

All the divisors of 10 are   1, 2, 5, 10,
All the divisors of 12 are   1, 2, 3, 4, 6, 12
But all the divisors of 11 are   1, 11

Every number is divisible by 1 and by itself. There are numbers which are not divisible by any other number besides these, such as for example 11. These are the numbers that are called prime numbers.

The number 1 behaves irregularly from this point of view, it has only one divisor, 1, and this is also itself. For this reason it is usual not to include 1 among the prime numbers. According to this convention the smallest prime number is 2, and this is also the only even prime number, as every even number is divisible by 2. In accordance with this an even number can only be a prime number if the divisor is the number itself, namely 2.

What gives prime numbers great importance is the fact that every other number can be built out of them as out of bricks. For this reason the other numbers are called composite numbers. We can formulate this more exactly by saying that every composite number can be expressed as a product of prime numbers.

Let us try for example to write down 60 as a product:

$$60 = 6 \times 10$$

Both 6 and 10 can be further split into factors,

$$6 = 2 \times 3 \text{ and } 10 = 2 \times 5$$

and writing these in place of 6 and 10 we have

$$60 = 2 \times 3 \times 2 \times 5$$

where all its factors are prime numbers.

We could have done this differently, since we have already seen that 60 can be expressed as a product of two numbers in a number of different ways. If we start with

$$60 = 4 \times 15$$

where $4 = 2 \times 2$ and $15 = 3 \times 5$, we have

$$60 = 2 \times 2 \times 3 \times 5$$

or if we choose the splitting

$$60 = 2 \times 30$$

then $30 = 5 \times 6$ and $6 = 2 \times 3$ so $30 = 5 \times 2 \times 3$

or $30 = 2 \times 15$ and $15 = 3 \times 5$ so $30 = 2 \times 3 \times 5$

or $30 = 3 \times 10$ and $10 = 2 \times 5$ so $30 = 3 \times 2 \times 5$

We see that 30 can always be split into the product of the prime numbers 2, 3, and 5. If we write this product in place of 30, we have

$$60 = 2 \times 2 \times 3 \times 5$$

Start off how we will, 60 will be split into the same prime numbers, only they may occur in different orders. Tidying up and writing the products of equal factors as powers, we have

$$60 = 2^2 \times 3 \times 5$$

It is just as easy to split any composite number into its prime factors (and it can be proved that every time we can obtain only one kind of split). If in the beginning we get stuck and do not know how to begin, let us remember that the smallest divisor of a number, apart from 1, is certain to be a prime number, since if it were composite, then there would have to be a divisor smaller than it, which would also go into the original number without remainder. So, by always looking for the smallest divisor, we can easily find the prime factors of any number. For example:

$$90 = 2 \times 45$$
$$= 2 \times 3 \times 15$$
$$= 2 \times 3 \times 3 \times 5$$

Such a splitting of a number throws light on the structure of the number; for example we can gather straight away that the divisors of 90, apart from 1, are

primes: 2, 3, 5

products of two primes:

$2 \times 3 = 6$, $2 \times 5 = 10$, $3 \times 3 = 9$, $3 \times 5 = 15$

products of three primes:

$2 \times 3 \times 3 = 18$, $2 \times 3 \times 5 = 30$, $3 \times 3 \times 5 = 45$

product of four primes: $2 \times 3 \times 3 \times 5 = 90$

So it is worth while to make friends with the bricks out of which numbers are built. Let us try to write down the prime numbers in order. We know that the smallest prime number is

2, and that we can certainly leave out all other even numbers, since all these are divisible by 2. Then 3, 5, 7 are
prime numbers; it is tempting to say that 9 is also a prime
number, but it is not, since 9 is divisible by 3. Now we might
think that from here on the prime numbers are going to thin
out, but again they do not, since 11 and 13 are prime numbers.
For once I shall ask the reader to take a little trouble: he
should try to list all the prime numbers for himself at least up to
50. As a check, the reader will find the series written below,
but he will only appreciate its irregularity if he has gone
through the series himself, having made a number of mistakes
while doing so.

The sequence of prime numbers is as follows:

2, 3, 5, 7, 11, 13, 17, 19, 23, 29, 31, 37, 41, 43, 47, . . .

A clever idea has been handed down to us from the
Greeks, by means of which we can construct this irregular
sequence mechanically without any possibility of error. Let
us write down all the numbers from 2 to 50. The first number
must be a prime number (even if we did not know what it was
we should be sure of this), as all its divisors apart from itself
must be smaller than itself, and (apart from 1) would have to
come before it, but there is nothing before it. Now let us see
what this first number is—2. Every second number is a
multiple of 2, and so, apart from 2 itself, is not a prime number.
So, starting from here, let us cross out every other number.

2, 3, ~~4~~, 5, ~~6~~, 7, ~~8~~, 9, ~~10~~, 11, ~~12~~,
13, ~~14~~, 15, ~~16~~, 17, ~~18~~, 19, ~~20~~, 21, ~~22~~, 23,
~~24~~, 25, ~~26~~, 27, ~~28~~, 29, ~~30~~, 31, ~~32~~, 33, ~~34~~,
35, ~~36~~, 37, ~~38~~, 39, ~~40~~, 41, ~~42~~, 43, ~~44~~, 45,
~~46~~, 47, ~~48~~, 49, ~~50~~

The first number which remains after the number 2 again can
only be a prime number, since it can only be a multiple of
numbers that come before it, and before it there is only a
number whose multiples we have crossed out. Let us have a

look at this number: 3.   Every third number is a multiple of 3,
so from here on let us cross out every third number (it does not
matter if some numbers get crossed out twice).

2, 3, ~~4~~, 5, ~~6~~, 7, ~~8~~, ~~9~~, ~~10~~, 11, ~~12~~,

13, ~~14~~, ~~15~~, ~~16~~, 17, ~~18~~, 19, ~~20~~, ~~21~~, ~~22~~, 23,

~~24~~, 25, ~~26~~, ~~27~~, ~~28~~, 29, ~~30~~, 31, ~~32~~, ~~33~~, ~~34~~,

35, ~~36~~, 37, ~~38~~, ~~39~~, ~~40~~, 41, ~~42~~, 43, ~~44~~, ~~45~~,

~~46~~, 47, ~~48~~, 49, ~~50~~

Then we go on in the same way.   We keep the number 5, but
we naturally cross out all multiples of 5.   So after 5 we must
cross out every fifth number, similarly after 7 we must cross out
every seventh number.

2, 3, ~~4~~, 5, ~~6~~, 7, ~~8~~, ~~9~~, ~~10~~, 11, ~~12~~,

13, ~~14~~, ~~15~~, ~~16~~, 17, ~~18~~, 19, ~~20~~, ~~21~~, ~~22~~, 23,

~~24~~, ~~25~~, ~~26~~, ~~27~~, ~~28~~, 29, ~~30~~, 31, ~~32~~, ~~33~~, ~~34~~,

~~35~~, ~~36~~, 37, ~~38~~, ~~39~~, ~~40~~, 41, ~~42~~, 43, ~~44~~, ~~45~~,

~~46~~, 47, ~~48~~, ~~49~~, ~~50~~

We need not go any farther, since the first remaining number is
11 and if we multiply 11 by a number greater than 7, we
obtain a number greater than 50, and the smaller multiples of
11 have already been crossed out.   Let us write out the numbers that have survived:

2, 3, 5, 7, 11, 13, 17, 19, 23, 29, 31, 37, 41, 43, 47

these are in fact the prime numbers under 50 which we wrote
down before.

   It is possible to build a machine which would carry out all
these instructions, and so would yield all prime numbers up to a
certain point.   But this does not alter the fact that prime
numbers turn up again and again in a most unpredictable
manner however far we like to go.

For example it can be proved that we can find gaps as large as we like between successive prime numbers, provided we go far enough in the number series. For example, the results of the following operations will yield a gap of six units, i.e. six successive numbers none of which is a prime number:

$$\underline{2} \times 3 \times 4 \times 5 \times 6 \times 7 + \underline{2}, \qquad 2 \times \underline{3} \times 4 \times 5 \times 6 \times 7 + \underline{3}$$

$$2 \times 3 \times \underline{4} \times 5 \times 6 \times 7 + \underline{4}, \qquad 2 \times 3 \times 4 \times \underline{5} \times 6 \times 7 + \underline{5}$$

$$2 \times 3 \times 4 \times 5 \times \underline{6} \times 7 + \underline{6}, \qquad 2 \times 3 \times 4 \times 5 \times 6 \times \underline{7} + \underline{7}$$

These are in fact successive numbers, each one being just one more than the previous one, and not one of them is a prime number, since $2 \times 3 \times 4 \times 5 \times 6 \times 7$ is divisible by each one of its factors. Therefore our first number is such that both terms are divisible by 2, the second number is similarly divisible by 3, the third by 4, the fourth by 5, the fifth by 6 and the sixth by 7. If we calculate this number, we have

$$2 \times 3 \times 4 \times 5 \times 6 \times 7 = 5040$$

so the six successive numbers are

$$5042, \; 5043, \; 5044, \; 5045, \; 5046, \; 5047$$

These numbers are quite large, and we have been obliged to go quite a long way in the number series to find a gap of six terms among the prime numbers by means of this method. It is of course possible that there is such a gap between them considerably earlier. If we are not averse to going a long way, we can find a gap of 100 terms in the same way, by adding to the product of all the numbers between 2 and 101

$$2 \times 3 \times 4 \times 5 \times \ldots \times 100 \times 101$$

first 2, then 3, then 4, and finally 101. By this method we can find as long a gap as we like.

On the other hand, as far as the number series has so far been examined, we find again and again successive odd numbers which have turned out to be prime numbers, as for example at the beginning of the number series we have 11 and 13 or 29 and 31. Mathematicians have an idea that there are in fact such 'twin' prime numbers however far we go in the number series, i.e. beyond the part so far examined: but so far it has not proved possible to show this in any general way.

Of course we might ask whether there are prime numbers among random numbers as large as we please? Do they just colour the first section of the number series? To this question we can provide an answer. We have in fact had the answer for 2000 years; Euclid published a very clever proof to show that there are an infinite number of prime numbers.

We can see this in much the same way as the infinity of the natural number series itself; should somebody say that the prime numbers end at such and such, he cannot get away with it, because we can show that there are still prime numbers beyond.

It is enough to show this in one case; it will be just the same in every other case. We need only remember that every other number is divisible by 2, every third number by 3 and so on. Therefore the number that comes immediately after a number divisible by 2 cannot itself be divisible by 2, the number that comes immediately after a number divisible by 3 cannot be divisible by 3 and so on. If somebody were to state that the following are all the prime numbers:

$$2, 3, 5, 7$$

and that here they end, we can refute this statement, since we can construct the following number out of the given ones:

$$2 \times 3 \times 5 \times 7 + 1$$

$2 \times 3 \times 5 \times 7$ is divisible by 2, by 3, by 5 and by 7. The number that comes immediately after this one, i.e. $2 \times 3 \times 5 \times 7 + 1$, cannot be divisible by any of these. But the poor thing must be divisible by some prime number, for it is still a number and so can be split up into prime numbers, or if by chance it is a prime number it can at any rate be divided by itself. The person who made the statement referred to must have made a mistake, there must be prime numbers beyond 7. And in the same way beyond any prime number.

Let us calculate this number

$$2 \times 3 \times 5 \times 7 + 1$$

The result is 211. We can convince ourselves after a few trials that this number is not divisible by any other number besides 1 or itself, i.e. it happens to be a prime number. So this is the very prime number whose existence I have stipulated, a prime number greater than 7. Of course it does not mean that this is the first prime number after 7; we could not have expected for one moment that the succession of prime numbers could be constructed in such a regular way.

More exactly stated, our method ensures that in order to find a prime number after 7 we certainly need not go beyond $2 \times 3 \times 5 \times 7 + 1$. In the same way to find one after 11 we need not go beyond $2 \times 3 \times 5 \times 7 \times 11 + 1$. But these are rather large distances; would it not be possible to find prime numbers within narrower limits?

Many people have given their attention to this problem. I shall just mention one beautiful result: a Russian mathematician Tchebicheff has proved that from the number 2 onwards there is always a prime number between any number and its double.

$$
\begin{array}{llll}
\text{Between 2 and} & \text{4 we have} & 3 \\
\text{between 3 and} & \text{6 we have} & 5 \\
\text{between 4 and} & \text{8 we have} & \text{5 and 7} \\
\text{and between 5 and 10 we have} & & \text{7 only}
\end{array}
$$

and although there does not appear to be any regularity in this, it is nevertheless true, however far we go in the number series.   We can find as many primes as we like between numbers and their doubles if we care to go far enough.

Thus we have found some regularity in these seemingly unmanageable prime numbers; they cannot get away from each other quite arbitrarily.

And in spite of everything, in a certain sense there is a 'rule of prime numbers'. It is in the sense of 'almost', the kind of sense in which a circle can be considered as being put together out of a lot of thin triangles (which I have promised to make precise later).

Up to and including 2 there is one prime number, namely 2 itself.   Up to 3 there are 2, namely 2 and 3, up to 4 again these same 2, up to 5 there are 3 since 5 is now added to the list, up to 6 it is still these 3, up to 7 there are already 4, i.e. 2, 3, 5, and 7.   Up to 8, 9, and 10 it is still these same 4, and so on. So the number of prime numbers is

up to 2	up to 3	up to 4	up to 5	up to 6	up to 7	up to 8	up to 9	up to 10
1	2	2	3	3	4	4	4	4

This sequence jumps every time we get to a prime number, and this happens at quite irregular intervals. Nevertheless we can write down a well-known sequence constructed according to a certain rule,* in which the farther we go the more they become like the numbers in our sequence, and so the numbers in these two sequences are 'almost' equal if we go far enough in both sequences, in the same sort of way as the curved subdivisions of the circle, so difficult to handle,

are more and more like the well-known triangles, and the longer we continue the subdivisions the more truly can the segments be said to be 'almost' the same as the triangles:

An exact rule is not even imaginable for prime numbers, but even this 'almost' kind of regularity has its exact meaning; I shall keep my promise and come back to this point later.

I could not even attempt to sketch the proof of the rule for prime numbers; the best mathematicians have handed it down over a long period of time, improving on it here and there, until it has reached its present form. Research is still in progress in this field, attempts are being made to gauge more and more accurately the extent of the error made by substituting for the terms of our irregular sequence those of the

---

* For the sake of those who still remember logarithms I give the sequence which is

$$\frac{2}{\log 2}, \frac{3}{\log 3}, \frac{4}{\log 4}, \frac{5}{\log 5} \cdots$$

It is unlikely that the reader will remember *this kind* of logarithm: it is the so-called natural logarithm. We shall come across it later on.

sequence that can be constructed according to a rule.   Here it is not utility that inspires research, nor convenience, but the beauty and the difficulty of the subject.   This is a very different kind of beauty from the beauty of our playful numbers in the results we saw in the theory of combinations; it is rather the aesthetics of lack of organization.   It is a noble task to constrain the irregular to conform to the regular.

The fact that there is a rule for prime numbers means that although the prime numbers are distributed irregularly along the sections of the number series that can be examined, they are nevertheless subject to some kind of order if they are considered in their infinite entirety.   I am reminded of a simile which I read somewhere in connexion with the problem of free will: if we observe a swarm of bees at close quarters they will appear to be flying hither and thither in all directions, although the whole swarm is nevertheless carried along in a certain direction towards a definite goal.

8. *'I have thought of a number'*

LET us go back for a while to the useful kind of Mathematics. We can already work out the volume of a cube; but often we need to know the volumes of other irregular-shaped solids, and we cannot work these out by means of direct measurements. In these cases we can make use of the following device: let us suppose that our solid is made of oak. We can weigh it. Then we can carve out of oak an inch cube (i.e. one whose volume is one cubic inch), and weigh it. The number of times the weight of the cube goes into the weight of the solid in question will be the volume of our solid in cubic inches.

Here we cannot determine volume directly, but we can determine something else, with which volume is in a well-known relationship, namely the weight of the solid. It is from this that we try to deduce the unknown volume.

This is a very common situation in Mathematics; a required quantity is unknown to us, but we do know certain relationships in which it stands to other quantities. From these relationships we may be able to find out the value of the unknown quantity.

From the point of view of applications such a process is fundamental, and it is essentially the same as the process of solving the following well-known type of problem: 'I have thought of a number, added something to it, multiplied the result by 3' and so on. Further operations are enumerated that are performed on the number in question, and eventually it is revealed that after all has been done the end-result is, for example, 36. The problem is, then, to find out what the original number was.

Well, let the reader find out: I have thought of a number, added 5 to it and got 7; what was the number I thought of? Everybody will know that it was 2.

Let us make it a little harder. I have thought of a number, multiplied it by 5, then divided it by 2, then added 3, and I obtained the number 18. What number did I think of?

Such problems are usually given verbally and not in writing, and so anyone trying to solve them soon forgets what the operations were; it is therefore better to jot them down as the problem is being given out.

Since we do not know the number, we call it X. If the poet Babits can write

> The Styx awaits all other rivers,
> Oh, X, the best of all resolvers!

then perhaps I shall be allowed to suggest that the resolver of the problem should write X in place of the unknown number awaiting discovery. He will then jot down the following: X was the original number, this was multiplied by 5, so it became $5X$, then it was divided by 2, so it became $\frac{5X}{2}$, then 3 was added, so it became $\frac{5X}{2} + 3$ and we know that this is 18.

In other words

$$\frac{5X}{2} + 3 = 18$$

The number originally thought of satisfies this kind of 'equation'; it is from this that its value can be found out.

There are people who have such feeling for numbers that they can find out the value of the number from the equation in this form. Those who cannot do this, can go back one step; if something became 18 after adding 3, then it must have been 15 before:

$$\frac{5X}{2} = 15$$

From this it is easier to find out what the X was. Those who still cannot do so, can make matters easier for themselves by going back another step. If we get 15 when we divide something by 2, then it must have been 30 before we divided it:

$$5X = 30$$

Now everybody will know that if you take a number five times and get 30, then the number can only be 6.

This gradual dismantling of an equation can be done with any other equation. When we passed from

$$\frac{5X}{2} + 3 = 18$$

to

$$\frac{5X}{2} = 15$$

the term 3 disappeared from the left and we subtracted 3 from the number on the right. This is what we mean when we say that a positive term can be taken over to the other side of an equation as a term to be subtracted. When

$$\frac{5X}{2} = 15 \text{ became } 5X = 30$$

the divisor 2 disappeared from the left, and we multiplied the number on the right by 2. This is expressed by saying that a divisor can be taken over to the other side of an equation as a multiplier. In general, opposite operations can be carried from one side of an equation to the other.

Even when confronted with a more cunningly devised equation, if we give the matter a little thought we shall see that this is still the same as the problem of finding the value of a number somebody thought of. Let the problem be formulated in the following way, for example: 'A father is 48 years old, his son is 23. In how many years will the father be just twice as old as the son?' Of course there are people who will find this out straight away without any equation. Those who are a little slower can think it out in the following way: the quick ones already know the result, they will have thought of the right number; for us this number is still X. So after X years the father will be twice as old as the son. How will those who have already solved the problem check their result? They will see how old the father will be in X years and how old the son will be in X years, and notice whether in fact the father will then be twice as old as the son. After X years the father's age is X more than 48, i.e. $48 + X$ years; the son's age is $23 + X$ years. So the clever ones thought of a number, added this number both to 48 and to 23 and they say that the result of the first addition is just double the result of the second:

$$48 + X = 2 \times (23 + X)$$

It is from this that we must find out what X is. The multiplication by 2 on the right can be carried out by multiplying both terms by 2:

$$48 + X = 46 + 2X$$

The X on the left can be taken over to the right as a subtraction, and the 46 on the right can be taken over to the left as a subtraction, so that all the X's accumulate on the same side of the equation

$$48 - 46 = 2X - X$$

$48 - 46 = 2$, and it is obvious if we take away one X from $2X$, we shall be left with one X:

$$2 = X$$

so the right number is 2. In 2 years' time the father will be twice as old as the son. In fact in 2 years the father will be 50 and the son 25.

Let us complicate things further. 'I have thought of two numbers, their sum is 10. What are the two numbers?'

We can write this as follows: let the two numbers be X and Y (if somebody's surname and Christian name are both unknown, he might be called XY). So the person who set the problem states that

$$X + Y = 10$$

It is easy enough to find such numbers. For example 1 and 9 will do. But also 2 and 8, or perhaps 4 and 6, might have been the numbers thought of, and there are still other solutions. Of course this is altogether unfair, for it is impossible to find out what the two numbers were from what we were given. If we really want to find out, we can justifiably ask: 'We want to know something more about these two numbers!' All right, then we can be told that the difference between the two numbers is 2:

$$Y - X = 2$$

Now we can find out what they are quite easily. The numbers whose sum is 10 and which differ from one another by 2 are 4 and 6.

Therefore in order to find out two unknowns, we need two equations, i.e. a so-called system of equations. If it is not immediately clear from these equations what the numbers are, we can use a few tricks in order to make them more easily accessible to us.

For example, if somebody had not discovered that the solution of the above system of equations is 4 and 6, he could have done the following: in the second equation carry the term to be subtracted from the left to the right as a term to be added, then Y will be left by itself:

$$Y = X + 2$$

We can see from this that the second number is 2 more than the first number. So we could formulate the problem more simply as follows: 'I thought of a number, then added a number to this which is 2 more than the number I thought of, and the result was 10; what number did I think of?' We can write this as follows:

$$X + (X + 2) = 10$$

and in this there is only one unknown. We can find out what this one unknown is, as we already know some tricks for doing this. And once we know what X is, we need not worry about what Y will be, since we know that it will be two more than X.

Here is another example: 'I have thought of two numbers. To the first I added double the second and got 11, and to double the first I added four times the second, the result being 22. What numbers did I think of?'

The problem can be written in a short form as follows:

$$X + 2Y = 11$$
$$2X + 4Y = 22$$

If the reader has any eyes, he should see straight away that the problem is not fair. Let us try: 1 and 5 satisfy the first equation, since

$$1 + 2 \times 5 = 11$$

and the same numbers satisfy the second equation too, since

$$2 \times 1 + 4 \times 5 = 22$$

So we might think that we had already found the unknown numbers. But let us have another look. The numbers 3 and 4 also satisfy the first equation, since

$$3 + 2 \times 4 = 11$$

as well as the second, since

$$2 \times 3 + 4 \times 4 = 22$$

It seems that every pair of numbers satisfying the first equation also satisfies the second. The second condition does not help in choosing among these pairs a certain definite one. But this is really quite natural. Whatever X and Y are, $2X$ is always twice as much as X and $4Y$ is always twice as much as $2Y$, so obviously their sum $2X + 4Y$ is twice as much as $X + 2Y$, so if $X + 2Y = 11$, then $2X + 4Y$ can only be 22. According to this, the second equation does not tell us anything new about the unknown numbers; it tells us just the same thing as the first, only in a more complicated way.

It would be even more unfair if we had to find the values of X and Y out of the system of equations

$$X + 2Y = 11$$
$$2X + 4Y = 23$$

We can rack our brains till doomsday; no two numbers will satisfy both the conditions. We have already seen that whatever X and Y are, $2X + 4Y$ is always the double of $X + 2Y$, so if $X + 2Y = 11$, then $2X + 4Y$ must be 22, it is quite impossible for it to be 23. The second condition contradicts the first.

To sum up: we can find out the values of two unknowns from two equations, provided these equations do not state just the same thing or they do not contradict each other.

Now how would we cope with a problem like the following? 'I have thought of a number, squared it, added to it 8 times the number I thought of, and got 9.' Writing it down:

$$X^2 + 8X = 9$$

Here there is only one unknown, but the additional complication is that it occurs raised to the second power. The equation is a 'quadratic' one.

But let us not begin with such a complicated quadratic equation. The simplest form is

$$X^2 = 16$$

Everybody can see in a flash that the number is 4, since 4 is the number whose square is 16.

The following is just as simple:

$$(X + 3)^2 = 16$$

since the number whose square is 16 is still 4, so here

$$X + 3 = 4$$

and from this everybody will see that $X = 1$.

In the last equation $(X + 3)^2$ occurred; let us remember how a sum of two terms is squared. To the square of the first term (here to X^2) we have to add double the product of the two terms (here $2 \times 3X = 6X$), and the square of the second term (here $3^2 = 9$). So our equation in expanded form would look like this:

$$X^2 + 6X + 9 = 16$$

But if we had been confronted with it in this form, we should have had no idea how to begin to solve it. Therefore we have to practise recognizing squares of sums of two terms even in their expanded forms. If for example we had the equation:

$$X^2 + 8X + 16 = 25$$

then we should need to notice that $8X = 2 \times 4X$, and that 16 is the square of the 4 occurring in the previous product, so that

$$X^2 + 8X + 16 = X^2 + 2 \times 4X + 4^2 = (X + 4)^2$$

and that therefore we are dealing with the equation

$$(X + 4)^2 = 25$$

which we can solve in the same way as the previous ones.

Of course we do not really alter the equation we have just been considering if we carry the 16 to the right as a term to be subtracted: $25 - 16 = 9$, giving us the equation

$$X^2 + 8X = 9$$

which is the form we started with. Even in this form we ought to be able to see that the left-hand side can be completed into the square of two terms:

$$X^2 + 8X = X^2 + 2 \times 4X$$

and the term that is missing to make $(X + 4)^2$ from this is $4^2 = 16$. If we add the same thing to both the left- and the right-hand sides of an equation, the two sides still remain equal; so let us here add 16 to both sides:

$$X^2 + 8X + 16 = 9 + 16$$
$$X^2 + 8X + 16 = 25$$

and we can cope with it in this form.

This completion into the square of two terms is always possible. If the quadratic term is not X^2 but, for example, $3X^2$ as in the following equation:

$$3X^2 + 24X = 27$$

then we can divide both sides of the equation by 3, since if the left-hand side and the right-hand side are equal, then their thirds will also be equal. The third of $3X^2$ is X^2, the third of $24X$ is $8X$ and the third of 27 is 9, so that

$$X^2 + 8X = 9$$

which equation we can solve in this form. If the numbers had not been all divisible by 3, or if the coefficient of X were an odd number, then fractions would come into the working; but for the time being I do not want to bother the reader with fractions or with subtractions, although they do not really represent any difficulties of principle.

In any case we see that we can carry out the completion into a perfect square, and in this form the equations can be solved.

The above methods of reasoning are typical of the way mathematicians think. Quite often they do not deliver a frontal attack against a given problem, but rather they shape it, transform it, until it is eventually changed into a problem that they have solved before. This is of course the good old convenient point of view which is made fun of by the following problem, well known in mathematical circles: 'There is a gas ring in front of you, a tap, a saucepan, and a match. You want to boil some water. What do you do?' The reply is usually given with an air of uncertainty: 'I light the gas, put some water in the saucepan and put it on the gas.' 'So far you are quite correct. Now I shall modify the problem: everything is the same as before, the only difference is that there is already enough water in the saucepan. Now what do you do?' Now the problem-solver speaks up, more sure of himself, knowing himself to be in the right: 'I light the gas and put the saucepan on.' Then comes the superior reply: 'Only a physicist would do that. A mathematician would pour the water away and say that he had reduced the problem to the previous one.'

Certainly this reduction is the essence of the solution of quadratic equations, not the formula which is derived from it, and which is learnt by pupils so effectively that they can recite it in their sleep years afterwards.

Another difficulty may arise; let us suppose that the completion of the perfect square has already been carried out on the left-hand side, but we cannot find a number whose square is equal to the number on the right-hand side. For example:

$$(X + 3)^2 = 2$$

If I really did think of a number and X stands for that number, then this could not happen, but among the more advanced applications of equations this situation could in fact arise. The problem is the inversion of the operation of raising to a power; we are looking for a number whose square is 2. This is called extracting the square root, and, as an inverse operation, belongs to a later chapter (where I shall also be dealing with the problem of how many solutions there could be to a quadratic equation—at the moment we think ourselves lucky to find one). But the reader need not worry: the problem can be solved.

Those who do not rely on the formula but understand the line of argument, can also solve some equations of a higher order if they are in certain forms. Given for example

$$(X + 1)^3 = 27$$

since $27 = 3 \times 3 \times 3 = 3^3$, so 3 is the number whose cube is 27. Therefore $X + 1 = 3$ and so $X = 2$.

We could expand $(X + 1)^3$ by means of the binomial theorem by now familiar to the reader; from such an expanded form we can see that it is obtained by cubing the sum of two terms, but the completion into a perfect cube is not possible with every cubic equation. But there are general processes for the solution of cubic equations, even for the solution of equations of the fourth degree; apart from the four fundamental operations and the extraction of the square root we have to use cube roots and fourth roots, i.e. find numbers whose cubes or whose fourth powers are equal to certain given numbers, for example to 2.

The branch of Mathematics dealing with equations is called Algebra. In the secondary school we used to call all those parts of Mathematics Algebra that did not belong to Geometry. It is certainly true that in every branch of Mathematics (even in Geometry) we keep coming across equations, so that students of Mathematics may have formed the idea that it is the study of equations, and Higher Mathematics is the study of more complicated equations. Certainly there was an epoch when most mathematicians directed their attention to Algebra, and they imagined the development of Mathematics, after the equations of the third and fourth degrees had been disposed of, would consist of finding clever methods of solving equations of the fifth, sixth and higher degrees. We can imagine the general consternation when Abel discovered the conditions which must be satisfied for an equation of any degree to be solved by means of the four fundamental operations and the extraction of roots; it was found that only the equations of the first, second, third and fourth degrees satisfy these conditions. It is quite out of the question for us to be able to solve, for example, equations of the fifth degree by means of our operations. It seemed that the algrebraists might as well down tools.

Here we reach one of the most romantic parts of the history of Mathematics. It happened that a young Frenchman of 20, Galois, fought a duel over a girl, and was killed in the duel. The night before he died he wrote a letter to a friend, and in this, as though in a last will and testament, put down his ideas which breathed new life into Algebra just at the time when it had all but lost its *raison d'être*.

Even though there is no general procedure for the solution of equations of the fifth degree, there are some equations of this type which we can solve. For example

$$X^5 = 32$$

and, similarly, $(X + 1)^5 = 32$

can very easily be solved: $2 \times 2 \times 2 \times 2 \times 2 = 32 = 2^5$. So the solution of the first equation is

$$X = 2$$

The solution of the second is obtained from

$$X + 1 = 2$$

and so it is

$$X = 1$$

But there are equations of quite different form which can also be solved. One such would, for example, be

$$X^5 + 2X^4 + X = 0$$

$X = 0$ is certainly a solution, since every multiple and every power of 0 is 0, so that $0^5 + 2 \times 0^4 + 0$ is in fact equal to 0.

This was the situation which enabled algebraic research to be resumed; even if we could not think of a general procedure, it was still an interesting problem to decide which were those special equations of higher degree which could be solved by means of our operations.

Galois's last will and testament gave a method of attack for this problem.

This method has proved remarkably fertile. It is due to it that Algebra, at the time quite moribund, started to flower again, with even greater force than before. Wherever Mathematics is wounded, regeneration invariably begins with renewed vigour. This new branch of Algebra enshrines Galois's memory in its very name; it is called the Galois Theory.

There is just one thing of which I should like the reader to be aware: it is in Algebra that we have first come across the phenomenon of Mathematics, by means of its own tools, proving its own incapacity within a certain circumscribed area. We shall come across this phenomenon again.

PART II

THE CREATIVE ROLE OF FORM

9. Diverging numbers

DURING the preceding chapters a whole heap of debts have accumulated; these have been in most cases in connexion with the inversion of operations. It is now time we faced the problems of the inverse operations.

Subtraction would appear to be the more harmless. What is it actually about? We can invert addition as follows: we know the sum of two terms, suppose it is 10. One of the terms is 6, what is the other? Of course it is 4. This is what is left if we take away the given term from 10; where is the difficulty?

The difficulty begins with the fact that I had to do a little thinking before deciding on the numbers 10 and 6 for our example. In the case of addition, I could point blindfolded to any two places in the natural-number series, and it would be certain that the numbers so found could be added, and what is more, that this could be done in any order. But what would have happened if I had formulated the above example by saying that the sum of two terms was 6, one term being 10; what would the other term be? Here even the statement of the problem clearly shows that the whole thing is impossible, since the sum cannot be less than one of its terms. We have to be careful not to subtract more than there is to subtract from.

Is this all? The reader will think it was a pity to leave this simple operation for so long just for this. It would not occur to anybody to take away more than there is, and other, sensible, subtractions can be carried out without any difficulty.

This would be quite a reasonable line to take but for the consideration that in fact problems arise in practice where we have to subtract a larger number from a smaller one.

Let us recall for a moment the problem where we had to find out in how many years the father would be twice as old as

77

the son and ask the same question about a father who is 52 and a son who is 27. The argument is the same as before: the situation in question will occur in X years' time, when everybody is X years older. The father will be $52 + X$ years old, the son $27 + X$ years old. We are saying that

$$52 + X = 2 \times (27 + X)$$

Let us do what we did before. On the right let us carry out both multiplications:

$$52 + X = 54 + 2X$$

Then let us collect the unknowns on the right-hand side, i.e. bring the X from the left over to the right as a subtraction, and take the 54 likewise over to the left:

$$52 - 54 = 2X - X$$

Taking one X from $2X$, only one X is left:

$$52 - 54 = X$$

but here we get stuck. The result of the impossible subtraction $52 - 54$ ought to be the value of the unknown number.

To this we could reply: if the unknown can only be the result of an impossible subtraction, then we must look for the mistake in the formulation of the problem itself; this father will never be twice as old as his son.

But let us have a closer look at the numbers occurring in the problem: 52 years and 27 years. Anyone who has some number sense will see straight away that 2 years *ago* the father was 50 and the son 25, and so that is when the father *was* twice as old as the son.

It seems that the example ought to be reformulated. How many years ago was the father twice as old as the son?

In this way we should have no trouble with the equation. X years ago everybody was X years younger, so the father was $52 - X$ years old and the son $27 - X$ years old, and we are stating about these ages that

$$52 - X = 2 \times (27 - X)$$

On the right we can multiply the difference by 2 by multiplying 27 as well as X by 2 (if we are confronted with the problem 2×99, it is easier to do it by multiplying 100 by 2, though from the 200 so obtained we must subtract 2, since we are considering 99 as the difference between 100 and 1, and so double this difference is the difference between 2×100 and 2×1):

$$52 - X = 54 - 2X$$

Now let us collect the unknowns on the left-hand side, i.e. let us bring the subtracted $2X$ over to the left as an added $2X$:

$$2X + 52 - X = 54$$

Now the added 52 can be taken over to the right as a subtracted 52:

$$2X - X = 54 - 52$$

We can now carry out all the subtractions and obtain

$$X = 2$$

just as we had thought.

This is all very well; we can always do this, but it is very tiresome. It means following the old patterns until we run up against a blank wall, then going back to the beginning, reformulating the problem, and starting the whole thing again. And all the time the solution was ready to hand. Let us go back to the point where we got stuck. The expression

$$52 - 54$$

itself solves the difficulty; it seems to say: 'I give away the fact that the difference is *2*! Moreover, I tell you that you need to look for these two years in the opposite direction, not ahead but in the past. Why can't you read this from me?'

Thus it is very tempting to attribute a meaning to the difference $52 - 54$. We ought to mean by it the same as the difference between 54 and 52, but it should somehow have a different direction, opposite to the usual one. Since this direction points towards the past in time, it indicates that we must subtract two years from today's ages: for it is usual to denote it by means of the sign for subtraction, and we usually write

$$52 - 54 = -2$$

Correspondingly, we ought to put $+$ signs in front of our numbers that we have dealt with so far, since if the result of solving our equation had been that the situation in question would occur in *two years' time*, then we should have had to add 2 years to today's ages. If we wish to emphasize this, we can put in the + sign.

The need to attribute directions to quantities is not uncommon. If on a very cold winter's day we say that the temperature outside is 4 degrees, we have not given any exact information about the temperature in the street. We need to add whether it is 4 degrees above or below zero; for a sensitive man this could represent quite a difference.

For the same reason it is loose to talk about the third century, without stating whether we mean A.D. or B.C.. or about 15 degrees longitude, without adding whether it is 15 degrees East or West of Greenwich. The bookkeeper is also careful whether an item of £10 is put on the left or on the right of the ledger, since for most people it would be a matter of some importance whether their account had increased or decreased by £10.

In all these cases we could use the signs '+' and '−' for the quantities which may have one of two directions. We give these signs special names: the '+' sign is called the positive sign, and the '−' sign is called the negative sign. The negative

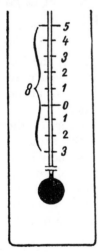

numbers can be thought of as results of subtractions in which we subtract a larger positive number from a smaller one.

If the temperature outside is 5 degrees above zero, and then it drops by 8 degrees, this means that it becomes less, and we can consider it as a subtraction; we must take 8 away from 5. There is no difficulty in this, we must merely go beyond zero; the temperature will be 3 degrees below zero, i.e. − 3 degrees.

$$5 - 8 = -3$$

Such a subtraction always takes us beyond the zero point in the direction opposite to the usual one. If we wish to represent our directed numbers on a line, we can show the positive numbers going one way (usually from left to right) and the negative numbers in the opposite direction

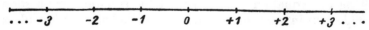

This very line can be thought of as one of our examples to illustrate the directed numbers. Let us imagine that it is a trunk road with a signpost placed at the zero point

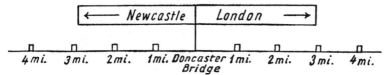

There are cases when we are interested only in the 'absolute' values of numbers, that is we are not interested in their directions. For example, we may want to know how far two points are from one another. The length of a snake, for example, is 3 yards, without any sign: for of course nobody thinks of it as 3 yards from head to tail, then 3 yards from tail to head, in all 6 yards.

It was really to be expected that opposites should turn up sooner or later in Mathematics; to think in opposites is so typically human: truth and falsehood, light and shade, thesis and antithesis.

The more exact mind will notice not just the most obvious of opposites. The transition from light to shade is infinitely gradual. From one starting point we can go in more than two directions: roads go in all directions from Doncaster

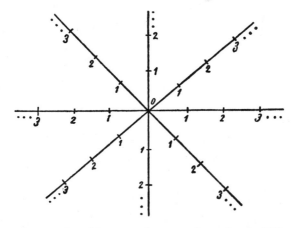

Bridge. So we should not only complete the half-line in our diagram on which we represented natural numbers with the

other half in the opposite direction, but with all possible lines radiating from the zero point.

This is not just abstract thinking; quantities with arbitrary directions, so-called 'vectors', play an important part in Physics. Motion can take place in any direction, a force can act in any direction. Even operations with such directed quantities have a meaning; it happens sometimes that two forces are acting at the same time and we are interested in the joint result. Every oarsman knows that if he wants to cross the river, he will not reach the other bank dead opposite his starting point but lower down, since he is not only being propelled by his own muscles, but also by the current

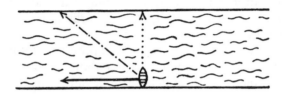

In still water the boat would proceed along the dotted line; if there were a current a boat that was not being rowed would be carried down along the thick line during the same time; the boat actually proceeds along the line it does under the influence of these two forces, and will get there as though it had done the two journeys separately one after the other:

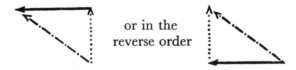

or in the
reverse order

In either order the boat would get to the same place. The terms are interchangeable even in this addition.

Even this kind of adding can be considered as successive counting. First we count the units of one vector in this direction ↑, then we go on counting in this direction ← with the units of the other vector; next we see what vector would have led us there direct. This is the sum, or, as it is called in this case, the resultant.

This is a queer sort of addition. For example and

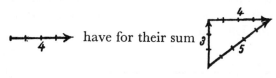 have for their sum

and, if we measure it exactly, we shall find that it consists of 5 units, and we obtain the apparently absurd result

$$3 + 4 = 5$$

But here we must not express ourselves loosely even for a moment; we must state what direction our 3 has, what direction our 4 has, and what direction our 5 has, and then it is not so impossible to get a sum less than $3 + 4 = 7$, since the result of opposite forces can even be zero. A story has it that eight horses were used to pull a very heavy cart, and the cart did not move at all, till somebody noticed that four horses were pulling one way and four the other way.

Let us leave the numbers diverging in all directions, and be content with just two opposing directions.

We already have our instructions how to add positive and negative numbers. For example if we want to add 8 and -5, then starting from zero we count 8 units to the right,

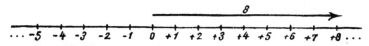

then 5 units to the left

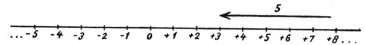

and we reach the number 3, so that the sum of $+8$ and -5 is $+3$. Let us not forget, on account of certain things that will come later, that

$$8 + (-5) = 3 = 8 - 5$$

so that instead of adding a negative number we can carry out a simple subtraction.

Subtraction is also quite easy on our line of numbers; it is the inverse of the previous process. For example, let us subtract -3 from 2. This means that a certain addition yields the sum 2, one of the terms is -3, and we are looking for the other term. In the case of addition we started in this way: we went 3 units to the left from zero

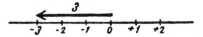

Now we are asking what we must have done to obtain $+2$ as the result? From -3 we have to go to the right in order to get to $+2$,

so that the difference between 2 and -3 is $+5$. This is quite interesting; it is just the same as if we had added 3 to 2. And it is always like this: instead of a subtraction, we can always do an addition, but we must add a number of opposite sign.

We might think that, since we know how to add, multiplication should not present any difficulty. To take -2 3 times means the following addition

$$(- 2) + (- 2) + (- 2)$$

and if from -2 we count another 2 to the left and then another, we get to -6, so

$$(+ 3) \times (- 2) = - 6$$

But supposing the multiplier is a negative number? We can add a number twice, three times or four times, one after the other, but there is no sense in which we can add them -2 times.

Now that we have a little experience, we shall not say lightheartedly: if it has no meaning, let us not do it; let us admit that we cannot multiply by a negative number. Negative numbers were introduced in order not to have to attack the same problem in two different ways, so as to have a uniform procedure in all cases. This is just the same with multiplication; if we are dealing with a problem which can be solved by multiplication by positive numbers it is most inconvenient to separate the cases and to say: if the numbers are positive, we multiply; if they are not, we do something else. Let us, rather, note carefully what this other thing is that we do in these cases, and let us call it multiplication by negative numbers. We are quite justified in doing this, since if something had no meaning before, we are at liberty to give it a meaning.

But an example will make matters clearer than a lot of talk.

If somebody is walking at a uniform rate of 3 miles an hour, how far will he have walked in 2 hours? The answer to this question is clearly given by a multiplication. If our walker walks 3 miles in one hour, then in two hours he will walk $2 \times 3 = 6$ miles. So I shall obtain the length of the walk by multiplying the length of time taken over the walk by the speed of the walker.

Now let us arrange the problem in such a way that the length of the walk, the time taken, as well as the speed, shall all be directed quantities. On the road there is a point which I shall call 'here'; walks to the right of this will be called positive, walks to the left will be called negative. If our walker does 3 miles an hour towards the right, his speed will be $+3$ miles per hour; if towards the left, his speed will now be called -3 miles per hour. Finally a moment in time can be chosen which can be called 'now', and the time after this can be regarded as positive and the time before it as negative. The starting point is always labelled 'Now the walker is here':

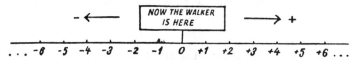

Let us restrict ourselves to the critical cases:

1. Somebody is walking at a speed of $+3$ miles an hour. Now he is here. Where was he two hours ago? Our result should be considered as the result of the multiplication:

$$(-2) \times (+3)$$

Let us think it over carefully: our walker has positive speed, so he is walking towards the right. Now he is here, he has arrived (the reader should point to the relevant point in the figure, i.e. where the notice is), so that 2 hours ago he must have been to the left of the notice. And he must have been at a distance to the left which he covered in 2 hours, i.e. at a distance of $2 \times 3 = 6$ miles. The point -6 is 6 miles to the left of the notice, so that

$$(-2) \times (+3) = -6$$

According to this, multiplying a positive number by a negative number, we obtain a negative number.

2. Now let the speed be -3 miles an hour. Our walker is here now. Where was he two hours ago? This can be regarded as the result of the multiplication

$$(-2) \times (-3)$$

The negative speed means that our walker is now walking towards the left, he has arrived here now (the reader should again point at the notice on the figure). This could have happened only if he was to the right of the notice two hours ago, and again 6 miles away. The point $+6$ is 6 miles to the right of the notice, so that

$$(-2) \times (-3) = +6$$

According to this the product of two negative numbers is positive.

This is rather like a double negation: 'It is not true that I have not been paying attention' means that I was paying attention.

From the rules of signs in the case of multiplication we can immediately deduce that the rule for division is similar. For example,

$$(+6) \div (-3)$$

means that we are looking for a number which if multiplied by -3 gives $+6$; this is of course obviously -2. The rule of signs for powers is just as easy:

$$(-2)^4 = \underline{(-2) \times (-2)} \times \underline{(-2) \times (-2)} = (+4) \times (+4)$$
$$= +16$$

and $(-2)^5 = \underline{(-2) \times (-2)} \times \underline{(-2) \times (-2)} \times (-2)$

$$= (+4) \times (+4) \times (-2) = (+16) \times (-2) = -32$$

In general, negative factors will yield positive products in pairs and we need only to see whether there is a factor left which is not in a pair.   In the same way an even power of a negative number is positive and an odd power is negative.

We arrived at the extension of the notion of multiplication from one single example; we might begin to wonder whether another example might not have led to different rules.   What alone can finally satisfy us on this point is the fact that the new rules of multiplication obey the same laws which we gathered from the old multiplication of natural numbers, and so we can use the rules without any fear of arriving at results contradicting the Mathematics that we had built so far.   For example here it is also true that the factors are interchangeable, since we saw that

$$(-2) + (-2) + (-2) = -6$$

i.e. $$(+3) \times (-2) = -6$$

From the example of the walker we have however obtained:

$$(-2) \times (+3) = -6$$

so that $$(+3) \times (-2) = (-2) \times (+3)$$

We must always be careful of this sort of thing.   Whenever we introduce new numbers or new operations, we should always see that they obey the old laws, since the reason we introduce them is to make procedures more uniform.   We do not want to have to split our operations into different types, depending on whether the new numbers or operations do or do not occur.   This regard for the extension of old, established concepts is called the 'principle of permanence'.

The natural-number series was a spontaneous creation.   It was the breaking down of the structure, which had worked quite well in the past, that led to the conscious creation of new numbers.   It is the shape of the structure which is helpful in this process.   The framework for the new numbers is given exactly by the laws derived from the old numbers, and we are unwilling to abandon these laws if we can help it.   This is the signpost in this conscious creation: we must shape the new numbers in such a way as to fit them well into this predetermined shape.   As Goethe says, since words give thoughts their shape:

> Where ideas are not yet,
> Words will serve to fill the net . . .

10. *Limitless density*

FOR the next problem we do not need even an equation; the smallest child can come across divisions which cannot be carried out with the natural numbers. Two children have to share an apple; they will be quite certain that neither of them can get a whole apple. They will quite simply cut the apple in half

 ½ an apple

and without any idea that by doing so they will have extended the concept of number.

So far we have regarded a unit as indivisible. Now we have introduced half of this as a new, smaller unit, and once we have taken this daring step, there is no reason why we should not divide a whole into 3, 4, 5 . . . or any number of parts, and then reckon with these new smaller units so obtained; for instance two halves, three halves, four halves, . . . or symbolically

$$\frac{2}{2}, \ \frac{3}{2}, \ \frac{4}{2}, \dots$$

This notation does not contradict the fact that we have denoted division by means of a line, since if for example we had to divide 2 into 3 parts, let us say three children are sharing out two cakes, the cleverest way of doing this is to divide both cakes into three parts, and every child then gets two-thirds, i.e. ⅔ of a cake.

The number under the line *names* the size of the unit, this is

the denominator.* The number over the line enumerates how
many of these units we have taken; this is the numerator.

In this way still more numbers have come into our line of
numbers. We can draw more and more lines, with smaller
and smaller units. A few will be found below:

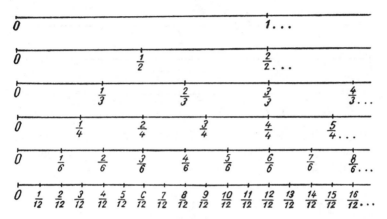

Among the numbers appearing on the different lines there
are some that are equal to one another. Let us see which are
the ones that come exactly below one another. Such numbers
are, for instance,

$$\frac{1}{2} = \frac{2}{4} = \frac{3}{6} = \frac{6}{12} \quad \text{or} \quad \frac{2}{3} = \frac{4}{6} = \frac{8}{12}$$

and from these we can work out what the apparent differences
are which do not actually alter the value of a fraction. For
example $\frac{4}{6}$ has the same value as the simpler-looking $\frac{2}{3}$; we can
'simplify' $\frac{4}{6}$ into $\frac{2}{3}$. We can carry this out by dividing both the
numerator and the denominator by 2: $4 \div 2 = 2$, $6 \div 2 = 3$,
$\frac{4}{6} = \frac{2}{3}$. But this is quite natural. Let us have a look; the
thirds are twice as big as the sixths. If, therefore, we take
half as much out of pieces that are twice as big, we get the
same amount.

We also see that $\frac{3}{3}$ is really one whole; $\frac{4}{3}$ on the other hand is
one whole and $\frac{1}{3}$ which is written for short as $1\frac{1}{3}$. These are not
really proper fractions at all, since their value is not a part of
the whole; they are called improper fractions.

* *nomen* is the Latin for name (*translator's note*).

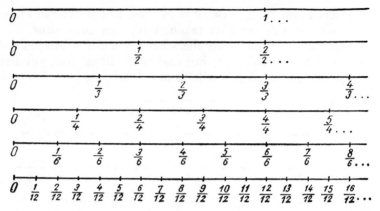

It is obvious that on one and the same line we can add and subtract by means of counting; for example from $\frac{3}{4}$ we get to $\frac{5}{4}$ by counting another two quarters, so that

$$\frac{3}{4} + \frac{2}{4} = \frac{5}{4}$$

Multiplication by a whole number is done in just the same way:

$$2 \times \frac{5}{12} = \frac{5}{12} + \frac{5}{12}$$

and counting five-twelfths beyond $\frac{5}{12}$ we get to $\frac{10}{12}$.

We get into a little difficulty when we have to add numbers that are counted in different sizes of units; for example we might want to do

$$\frac{2}{3} + \frac{3}{4}$$

We can get round the problem in the following way: let us find the line which has on it a number equal to both $\frac{2}{3}$ and $\frac{3}{4}$ (we need to do only a little thinking to realize that there will always be such a line). Here the line of the twelfths will do:

$$\frac{8}{12} = \frac{2}{3} \quad \text{and} \quad \frac{9}{12} = \frac{3}{4}$$

and now we can carry out the addition

$$\frac{8}{12} + \frac{9}{12}$$

on the same line.

We have similarly to leap across to another line if we want to divide.* The reader should check by measurement that the half of $\frac{1}{2}$ is $\frac{1}{4}$ and that the quarter of $\frac{2}{3}$ is equal to $\frac{2}{12}$. But this is to be expected, since a denominator which is four times as large means that we have divided something into four times as many pieces, and we have taken the same number of these pieces as we took of the larger pieces. In this way what we get is one-fourth of what we had before: for example, if we are dealing with cakes

So we see that, if we apply any of our fundamental operations to fractions, we again get fractions as a result, either proper or improper; it does not matter that, while working on them, the stops of the organ we play on must at times come from different rows.

We only have to face a real problem when we come up against multiplication by a fraction. There is no meaning in the statement: add something repeatedly a half times. A pupil of mine once said to me: 'If one whole times 3 is 3, then $\frac{1}{2}$ times 3 is 3,' and there was of course something in what he said. However, everyday speech helps us. 'Peter is $\frac{2}{3}$ times the height that his brother is' means that Peter's height is two-thirds of the height of his brother. To take something $\frac{2}{3}$ times means not taking the whole thing but just two-thirds of it. And this is that multiplication which recommends itself in, e.g., the following example: If a pound of tea costs 5 shillings, then 4 lbs. will cost $4 \times 5 = 20$ shillings, so that we obtain the cost by multiplying the price of one pound by the number of pounds that we buy. Modifying the problem: the price of a pound of tea is 5 shillings, how much is $\frac{3}{4}$ lb.? The result that we get ought to be called the result of the multiplication

$$\frac{3}{4} \times 5$$

---

* Here I am reminded of the fact that, during radiation, electrons leap across from one possible path to another. Perhaps some readers will find this comparison from atomic theory meaningful.

The price of $\frac{3}{4}$ lb. is obviously 3 times the price of $\frac{1}{4}$ lb. The price of $\frac{1}{4}$ lb. will be a quarter of 5 shillings (this is 1s. 3d.). This must be multiplied by 3 (this will be 3s. 9d.), so that $\frac{3}{4} \times 5$ does really mean that we take $\frac{3}{4}$ of 5. This can of course be carried out by dividing by 4 and multiplying by 3.

Quite similar considerations lead to the requirement that to divide by $\frac{3}{4}$ we must multiply by 4 and divide by 3. So these operations again give as results other fractions, on one or other of the lines, and it can be shown that in spite of the extension of the concept of multiplication, all previous operational rules remain intact. We must not be surprised that it can happen that the product may be smaller than the number we multiply. To multiply a number by $\frac{2}{3}$ means taking a $\frac{2}{3}$ part of it, and this is obviously smaller than the number itself.

It is very easy to multiply 20 by $\frac{1}{4}$; we simply take one quarter of 20, which is 5. It is just as easy to multiply it by $\frac{1}{2}$, by $\frac{1}{3}$ and by $\frac{1}{5}$, since in this case we must merely take the half, the third or the fifth of 20. For this reason it is sometimes worth while to split fractions into such 'partial' fractions,

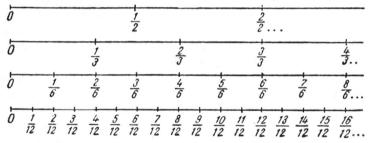

for example

$$\frac{5}{12} = \frac{4}{12} + \frac{1}{12}$$

Let us verify on the relevant lines that $\frac{4}{12}$ is the same as $\frac{1}{3}$, so that

$$\frac{5}{12} = \frac{1}{3} + \frac{1}{12}$$

and $\frac{1}{12}$ is one quarter of $\frac{1}{3}$ (the reader should look at the figure and see). According to this, for example,

$$84 \times \frac{5}{12} = 84 \times \left(\frac{1}{3} + \frac{1}{12}\right)$$

and this multiplication can be carried out by taking the third of 84, that is 28, then taking a quarter of this, that is 7, and 28 + 7 is 35. This is particularly useful in Britain where the relics of all sorts of number systems are preserved in the units of measurement. For example, the British shilling is split into 12 pence, and so in Britain multiplications by twelfths are very common.

We have seen that all our fundamental operations can be carried out within the field of fractions. Let us just do another example to show this: somebody is doing some arithmetic exercises; he does the easiest ones in $\frac{1}{3}$ of an hour (i.e. in 20 minutes), and the hardest ones in $\frac{1}{2}$ an hour. What is the average time he spends on an exercise? On the hardest and on the easiest exercise he spends altogether

$$\frac{1}{3} + \frac{1}{2}$$

hours. If these were equally difficult, half of this sum would be spent on one exercise. He probably spends about that time on an exercise of average difficulty. Let us calculate how much this time is. On the line occupied by the sixths we find numbers equal to both $\frac{1}{3}$ and to $\frac{1}{2}$,

i.e. $\qquad \frac{2}{6} = \frac{1}{3}$ and $\frac{3}{6} = \frac{1}{2}$

so that the sum of $\frac{1}{3}$ and $\frac{1}{2}$ is the same as

$$\frac{2}{6} + \frac{3}{6} = \frac{5}{6}$$

Half of this (please have a look) is the same as the number on the line for the twelfths

$$\frac{5}{12}$$

so that an average exercise takes $\frac{5}{12}$ of an hour (25 minutes). This is of course more than he would spend on the easiest and less than he would spend on the hardest exercises.

We can calculate the average of any two numbers by taking half of their sum. In this way we always obtain a number whose value lies between the two given numbers. That is why mathematicians call it the arithmetic mean.*

* Middle (*translator's note*).

This seemingly innocent example opens up enormous possibilities if we spend just a little time thinking about it.

First of all let us unite all our lines on to one line. There is no reason why we should not do this, why we should not represent all fractions on one line. It was just easier in the beginning to have separate lines for the different units, since on the unified line all the fractions whose value is the same will coincide at one point. Below I shall write every fraction in the form in which it occurred first:

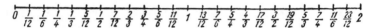

This is already becoming quite dense, but let us remember that we have united only a few lines: the fifths, the sevenths, the thirteenths, the hundredths and so on, an enormous number of subdivisions, do not figure on our line. Let us try somehow to find our way amongst them.

First of all we see that the whole numbers are amongst them. These can be considered as fractions whose denominators are 1. For example $\frac{3}{1}$ is really 3, if we think of that meaning of fractions which tells us that this is 3 divided by 1. The whole numbers and the fractions together are called rational numbers, and this foreshadows the possibility of numbers that are constructed in a less rational way.

Apart from zero (which can be considered as $\frac{0}{2}$, or $\frac{0}{3}$ or $\frac{0}{4}$ and so on) which will be the smallest fraction? It is obvious that $\frac{1}{12}$ is not the smallest, since $\frac{1}{13}$ is still smaller. If we divide a cake into one more number of equal pieces than before, then the pieces will be smaller. But the same thing is true whichever fraction we try. $\frac{1}{101}$ is smaller than $\frac{1}{100}$, $\frac{1}{1001}$ is smaller than $\frac{1}{1000}$. So among the rational numbers there is not only no greatest, there is no smallest either.

We cannot therefore start the enumeration of the rational numbers. So let us start with any small fraction, for example with $\frac{1}{12}$, and try to enumerate all the rational numbers at least from this point onwards. What is the fraction that comes after $\frac{1}{12}$? This cannot be the $\frac{1}{6}$ which follows it on our line, since we know that the arithmetic mean of $\frac{1}{12}$ and $\frac{1}{6}$ comes between these two fractions. But even if we had chosen some other number to the right of $\frac{1}{12}$ instead of $\frac{1}{6}$, we could have constructed the arithmetic mean of $\frac{1}{12}$ and of this number, and this again would have been nearer to $\frac{1}{12}$ than the number we thought of. So there is no number which comes immediately after $\frac{1}{12}$. It is therefore equally impossible to enumerate the rational numbers starting with $\frac{1}{12}$. We can see that, in quite a general way, if we choose any two rational numbers, however close to one another on our line, these two numbers will not be immediate neighbours, for there will always be another rational number between them. This is what we mean when we say that the set of rational numbers is 'everywhere dense'.

Here we have come across another aspect of the Infinite; hard upon the limitless growth of the natural numbers and of prime numbers, comes limitless density. There is no number so large that there is not a larger one among the sequence of natural numbers or among the sequence of prime numbers. This is the exact meaning of what mathematicians say when they state that these sequences tend to Infinity. There is no distance so small that within this distance of $\frac{1}{12}$ there are no other rational numbers. This is expressed by saying that $\frac{1}{12}$ is a condensation point of the set of rational numbers. Of course not only $\frac{1}{12}$; all other rational numbers are condensation points.

In spite of the foregoing, it is still possible to arrange all the rational numbers in one sequence, but not in order of size.

We have already seen that we can arrange them in an infinite number of sequences, when we represented all fractions with the same denominators on one line. For the sake of uniformity let us also write the whole numbers in the form of fractions:

$$\frac{1}{1}, \frac{2}{1}, \frac{3}{1}, \frac{4}{1}, \frac{5}{1} \cdots$$

$$\frac{1}{2}, \frac{2}{2}, \frac{3}{2}, \frac{4}{2}, \frac{5}{2}, \cdots$$

$$\frac{1}{3}, \frac{2}{3}, \frac{3}{3}, \frac{4}{3}, \frac{5}{3} \cdots$$

$$\frac{1}{4}, \frac{2}{4}, \frac{3}{4}, \frac{4}{4}, \frac{5}{4}, \cdots$$

$$\frac{1}{5}, \frac{2}{5}, \frac{3}{5}, \frac{4}{5}, \frac{5}{5} \cdots$$

and so on. Now we must rearrange these in one single sequence. This can be done by writing down the numbers one after the other along the slanting lines that have been drawn in. In this way every one of them will have its turn:

$$\underbrace{\frac{1}{1}}, \ \underbrace{\frac{2}{1}, \frac{1}{2}}, \ \underbrace{\frac{3}{1}, \frac{2}{2}, \frac{1}{3}}, \ \underbrace{\frac{4}{1}, \frac{3}{2}, \frac{2}{3}, \frac{1}{4}}, \ \underbrace{\frac{5}{1}, \frac{4}{2}, \frac{3}{3}, \frac{2}{4}, \frac{1}{5}}, \cdots$$

and we shall have successive groups, getting longer and longer, but each one consisting of a finite number of numbers. Thus we do in fact get one single sequence, and anyone could continue this sequence once he has understood its rule of construction. He need not even look at the slanting lines in the above figure if he notices that the sum of the numerator and the denominator of the single fraction in the first group is 2, in both fractions in the second group the sum of the numerator and the denominator is 3; this sum is 4 in the fractions of the third group; in the group which was the last to be written down this sum is 6. On this basis

$$7 = 6 + 1 = 5 + 2 = 4 + 3 = 3 + 4 = 2 + 5 = 1 + 6$$

and so the next group can be constructed as follows:

$$\frac{6}{1}, \frac{5}{2}, \frac{4}{3}, \frac{3}{4}, \frac{2}{5}, \frac{1}{6}$$

Now anyone could continue the process mechanically. An infinite sequence is usually regarded as completely given if after recognizing its rule of construction anybody can write down its numbers to any desired point.

In our sequence there will of course be numbers whose values are the same; we already saw this with our lines. If we want to write every rational number only once, then we must add to the rule for their construction that those fractions that can be simplified should be left out. For example from the part of the sequence that has been written down we must leave out $\frac{2}{2}, \frac{4}{2}, \frac{3}{3}, \frac{2}{4}$. Among these $\frac{2}{2}$ and $\frac{3}{3}$ have the same value as $\frac{1}{1}$, $\frac{4}{2}$ the same as $\frac{2}{1}$, and $\frac{2}{4}$ the same as $\frac{1}{2}$. So in fact the sequence of rational numbers begins as follows:

$$\frac{1}{1}, \frac{2}{1}, \frac{1}{2}, \frac{3}{1}, \frac{1}{3}, \frac{4}{1}, \frac{3}{2}, \frac{2}{3}, \frac{1}{4}, \frac{5}{1}, \frac{1}{5}, \cdots$$

and this can be continued mechanically. I can state successively what the first, second, third . . . numbers are in the sequence. The sequence can be numbered, counted. This is described by a slightly misleading technical term as 'enumerable', or 'countable'.

This simple fact again sheds light on another surprising state of affairs. This is that, in spite of the fact that the rational numbers (i.e. all fractions) are everywhere dense, in some sense there are just as many rational numbers as there are whole numbers. How can we compare infinite sets with one another? A very simple method suggests itself for doing this. If in a dancing class I want to know whether there are just as many boys as girls present, I do not have to count them all. It is enough to tell the boys to take their partners. If after this no boy remains without a partner, and there are no wallflowers left, then it is clear that there are the same number of boys as of girls. This comparison can be applied to infinite sets; if we can pair the elements of two infinite sets in such a way that no element remains without a pair from the other set, then we say that these sets are equally numerous.

Now the sequence of rational numbers we have just constructed can be paired with the sequence of natural numbers. Let us pair 1 with $\frac{1}{1}$, 2 with $\frac{2}{1}$, 3 with $\frac{1}{2}$, so, for example, 10 will be paired with the 10th number in our sequence, i.e. with $\frac{4}{1}$; if we want to know what will be the number with which 100 is paired, we have only to construct the 100th number in our sequence by means of the given procedure, and that will be the required number. It is obvious that anybody could carry on with this pairing to any desired point, and it would be impossible to give a number, either out of the sequence of rational numbers or out of the sequence of natural numbers, that would be left without a partner. In this sense the set of natural numbers is as numerous as the set of rational numbers, and this is so in spite of the fact that the natural numbers can be imagined as scattered like raisins in a cake within the everywhere dense set of rational numbers, forming an apparently negligible minority of these.

This again throws light on a very important state of affairs; we must treat Infinity very gingerly. There are those who believe that it is a logical principle of permanent validity that the part is smaller than the whole. Here we have just seen an example of the contrary; the natural numbers form only a negligible part of the set of rational numbers, and yet they are just as numerous as the rational numbers. Such general logical principles have been abstracted from a great variety of human experience, but all such experience is taken from the finite. It has already led to much confusion when principles derived from finite experiences have been used to clothe the infinite. The infinite soon shakes itself free of such unsuitable clothing.

The fact that, in spite of everything, we still seem to kick against the possibility of the part becoming equal to the whole in any situation whatever, is probably due to unconscious forces, apart from experience itself, lending support to logical principles. The very foundations of morality itself seem to be undermined if the part can start competing with the whole. But perhaps for this reason there might be something of the joys of forbidden fruit in venturing out of the world of inexorable laws into the freedom of Infinity.

11. *We catch Infinity again*

LET us come back for a little while from the Infinite to the tangible world, and consider again the fact that on our hands, with which we try to reach out into this world, there are 10 fingers. Would it not be possible to force fractions into the decimal system too?

Let us remember what this system was: to the left of the ones was the place for units ten times as big, i.e. for the tens. To the left of these was the place for the hundreds, which are again ten times as big and so on. The idea seems to suggest itself that we continue this arrangement to the right as well: let us write the tenths in the first place to the right of the ones, in the second place let us write the tenths of tenths, i.e. the hundredths, in the third place the thousandths and so on. But we must somehow separate these new units from the ones, for although I might intend that in

$$1 \; 2$$

the 1 means a one, and the 2, two-tenths, everybody else would read it as twelve. That is why the decimal point is used:

$$1 \cdot 2$$

and it must not be forgotten that this is merely an abbreviation for

$$1 + \frac{2}{10}$$

In just the same way

$$32 \cdot 456 = 32 + \frac{4}{10} + \frac{5}{100} + \frac{6}{1000}$$

and this is the way we obtain decimal fractions or decimals.

Those fractions, whose denominators are 10, 100, 1000 or any other unit of the decimal system, can all be written down in decimal form too. For instance

$$\frac{23}{100} = \frac{20}{100} + \frac{3}{100}$$

But in $\frac{20}{100}$ we can divide both the numerator and the denominator by 10, so that

$$\frac{23}{100} = \frac{2}{10} + \frac{3}{100}$$

and since there are no whole numbers in this, we have finally

$$\frac{23}{100} = 0 \cdot 23$$

We might wonder whether every fraction can be written down as a decimal?

The simplest method of transformation is to carry out the division indicated in the fraction.

$$\frac{6}{5} = 6 \div 5 = 1, \text{ remainder 1.}$$

The remaining 1 can also be changed into tenths, so that it becomes 10 tenths, and dividing this by 5 we get 2 tenths. In the answer we must put in the decimal point:

$$6 \div 5 = 1 \cdot 2$$

so that $\qquad\qquad \frac{6}{5} = 1.2.$

Similarly $\qquad \frac{7}{25} = 7 \div 25 = 0.2$

and we are left with 20 tenths; these we can change into 200 hundredths, and if we divide this by 25 we get 8 hundredths, so that

$$7 \div 25 = 0.28$$

and $\qquad\qquad \frac{7}{25} = 0.28.$

But quite often we get stuck in the simplest cases:

$$\frac{4}{9} = 4 \div 9 = 0 \cdot 44 \ldots$$
$$40$$
$$40$$
$$4$$

and this division will never end. We can carry it on as long as we like but we always have a remainder 4. So $\frac{4}{9}$ cannot be written down as a decimal.

But how convenient it is to reckon with decimals! Let us just quote one example: it is absolutely child's play to multiply a decimal by 10. For example, if we have to do the following:

$$45 \cdot 365 \times 10$$

then we merely have to remember that 10 times 4 tens will be 4 hundreds, 10 times 5 units will be 5 tens, ten times 3 tenths will be 3 wholes (ones) and so on. We see immediately that we can carry out the whole multiplication by merely shifting the decimal point one place to the right:

$$453 \cdot 65$$

since in this way every place-value has moved to the left, with, for example, tens becoming hundreds. If we multiply our result by another 10, we get

$$4536 \cdot 5$$

giving us one hundred times the original number (for example the 5 ones have become 5 hundreds), and so we can see immediately that to multiply by 100 we must shift the decimal point two places to the right. In the same way we see that division by 10 can be carried out by shifting the decimal point one place to the left. This really is very little trouble. How useful it would be if we could write all fractions as decimals!

Now let us have a look again at where we got stuck before.

$$\frac{4}{9} = 4 \div 9 = 0 \cdot 44 \ldots$$
$$40$$
$$40$$
$$4$$

The remainder here will always be 4; this will in turn become 40 when we change down to units which are one-tenth of the previous units, and into 40 9 will always go 4 times. Even if this division is never finished, we still have the answer at hand: the 4 is going to repeat itself indefinitely.

The practical man will say to this: Even if the division came to an end at the tenth place, I should not make use of the whole answer; I am only interested in decilitres at most (one decilitre is one tenth of a litre), or in centimetres (one centimetre is one hundredth of a metre), or perhaps in grammes (one gramme is one thousandth of a kilogramme). The negligible amount that is left after one thousandth it would be real hairsplitting to consider at all. I need only the following from the infinite decimal:

$$0 \cdot 4$$
or $$0 \cdot 44$$
or $$0 \cdot 444$$

so I can do all my calculations with $\frac{4}{9}$ in the same way as with an honest-to-goodness finite decimal.

Physicists might need more figures than these in their much more accurate measurements, but even in these there is the so-called margin of error. A physicist can estimate, if an experiment is to be repeated, what sort of variations can be expected on account of the inaccuracies in our own senses and in the instruments themselves, and decimal places beyond this point are not worth bringing into the calculations. It can be assumed that instruments will be perfected more and more, that the margin of error will be reduced, but some error will always remain, and we shall always be able to stop somewhere in the sequence

$$0.4444444 \ldots$$

of decimal places, although possibly at a very distant place. It does not matter that we do not know in advance how far we shall have to go in the distant future; we certainly know that we shall be able to go to the place required, since we know the expansion of $\frac{4}{9}$ beyond any possible limit. However far we go in the expansion, the figure 4 will only repeat itself.

Is it then possible to transform every fraction into a decimal at least in this sense? Or, putting the question in another way, although a division is never finished, do the numbers in the answer at least follow one another according to some rule, which would enable us to get a general idea of the whole expansion?

We can easily see that the answer to this is in the affirmative. Every such expansion will sooner or later begin to repeat itself. Let us, for example, examine the fraction $\frac{21}{22}$:

If we divide by 22, the remainder is bound to be less than 22. If the division is never finished, every remainder will be one of the numbers:

1, 2, 3, 4, 5, 6, 7, 8, 9, 10, 11, 12, 13, 14, 15, 16, 17, 18, 19, 20, 21

Let us suppose that we have a chest of drawers with 21 drawers in it, and these numbers are written one on each drawer. If, while we carry out the division, we get a remainder 7, let us put a ball into drawer number 7. If we carry on patiently with the division until we have completed 22 steps, we have had to put

22 balls into 21 drawers, and so there is bound to be at least one drawer with two balls in it. After 21 steps one of the remainders must be repeated. If we are lucky, one of them will get repeated well before this and, once one of the remainders turns out the same as a previous one, everything will repeat itself from here on. Let us see this in our example:

$$\frac{21}{22} = 21 \div 22 = 0.954$$

$$210$$
$$120$$
$$100$$
$$12$$

Stop! We have already had 12 as a remainder. Here the numbers begin to repeat themselves:

$$21 \div 22 = 0.9545454 \ldots$$

$$210$$
$$120$$
$$100$$
$$120$$
$$100$$
$$120$$
$$100$$

so that, apart from an 'irregular' 9, 54 gets repeated indefinitely.

Conversely, if we are presented with such a periodic decimal, is it possible to find out of what fraction it is the expansion? Let us begin with 0·9545454 . . . and let us suppose that we do not know the fraction of which this is the expansion. Since we do not know it, we shall call it X:

$$X = 0.9545454 \ldots$$

If we multiply this by 1000, i.e. we carry the decimal point three places to the right, the whole-number part will be represented by the part of the expansion as far as the end of the first period,

$$1000X = 954.545454 \ldots$$

If, on the other hand, we multiply X by 10, the whole-number part will be the irregular part before the periods begin:

$$10X = 9.545454 \ldots$$

If we subtract the latter from the former, we must take 10 times

X from 1000 times *X* and so we are left with 990 times *X*. On the other hand the part after the decimal point consists in the case of both numbers of the indefinite repetition of 54, and so these parts are quite identical; when we subtract them, they will cancel out. The difference between 954 and 9 is of course 945, so that finally we have

$$990X = 945$$

Let us take 990 over to the right as a divisor; we shall have

$$X = \frac{945}{990}$$

This fraction can be simplified by 45:

$$945 \div 45 = 21 \quad \text{and} \quad 990 \div 45 = 22$$

so that

$$X = \frac{21}{22}$$

as of course we already knew.

During the working we made one careless step: we were not careful with the Infinite. We imagined 0·9545454 ... not just up to a certain degree of accuracy, but written down indefinitely, and we multiplied it as though it was a common-or-garden finite number. How are we justified in thinking that 0·9545454 ... has any sort of finite meaning whatever?

Let us consider this problem in a simpler example, since it is just as dubious whether

$$1·11111111111 \ldots$$

where the 1's are repeated indefinitely, has any finite meaning. It is a curious thing that people on the whole do not boggle over an infinite decimal of this kind, but they look aghast at an infinite addition like this:

$$1 + \frac{1}{10} + \frac{1}{100} + \frac{1}{1000} + \frac{1}{10000} + \ldots \text{ ad infinitum}$$

although this is just another way of writing the other. But I am not surprised at their looking aghast at the latter, though rather surprised that they accept the former. The sequence

$$1, \quad \frac{1}{10}, \quad \frac{1}{100}, \quad \frac{1}{1000}, \quad \frac{1}{10000}, \ldots$$

can be regarded as given even in its infinite extent, since anyone could continue it as far as he likes, but to regard this infinite path as so well-trodden that we can add all its terms is surely a little daring to imagine.   What can we understand by it?

A well-known mathematician while still a child formulated for himself the meaning of the sum of an infinite series in the following way.

There was a type of chocolate which the manufacturers were trying to popularize by putting a coupon in the silver-paper wrapping, and anyone who could produce 10 such coupons would get another bar of chocolate in exchange.   If we have such a bar of chocolate, what is it really worth?

Of course it is worth more than just one bar of chocolate, because there is a coupon in it, and for each coupon you can get $\frac{1}{10}$ of a bar of chocolate (since for 10 you can obtain one bar of chocolate).   But with this $\frac{1}{10}$th of a bar will go one-tenth of a coupon, and if for one coupon we get $\frac{1}{10}$th of a bar of chocolate, for $\frac{1}{10}$th of a coupon we get one tenth of this, i.e. $\frac{1}{100}$th of a bar of chocolate.   To this $\frac{1}{100}$th of a bar of chocolate belongs $\frac{1}{100}$th of a coupon, and for this we again get one-tenth as much chocolate, and one-tenth of $\frac{1}{100}$ is $\frac{1}{1000}$th of a bar of chocolate, and so on indefinitely.   It is obvious that this will never stop, so that my one bar of chocolate together with its coupon is in fact worth

$$1 + \frac{1}{10} + \frac{1}{100} + \frac{1}{1000} + \ldots \text{ bars of chocolate}$$

On the other hand, we can show that this is worth exactly $1\frac{1}{9}$ of a bar of chocolate.

The 1 in this is of course the value of the actual chocolate, so all that needs to be shown is that the coupon that goes with it is worth $\frac{1}{9}$th of a bar of chocolate.   It is enough to demonstrate that 9 coupons are worth one bar of chocolate, since then it is certain that one coupon is worth $\frac{1}{9}$th of this.   Suppose that I have 9 coupons, then I can go into the shop and say: 'Please can I have a bar of chocolate?   I should like to eat it here and now and I will pay afterwards.'   I eat the chocolate, take out the accompanying coupon, and now I have 10 coupons, with which in fact I can actually pay and the whole business is concluded, I have eaten the chocolate and I have no coupons left. So the exact value of 9 coupons is in fact one bar of chocolate, the value of one coupon is $\frac{1}{9}$th of a bar of chocolate, one bar of chocolate with a coupon is worth $1\frac{1}{9}$ bars of chocolate. So the sum of the infinite series

$$1 + \frac{1}{10} + \frac{1}{100} + \frac{1}{1000} + \frac{1}{10000} + \cdots$$

is exactly $1\frac{1}{9}$, quite tangibly, even edibly.*

We can sum up the result roughly as follows: if something is equal to 1 as a first rough approximation, equal to $1 + \frac{1}{10}$ as a slightly better approximation, equal to $1 + \frac{1}{10} + \frac{1}{100}$ as a still better approximation but still not exact, and so on indefinitely, then it is equal to $1\frac{1}{9}$ exactly, not approximately.†

Now I can keep the promises I made in the previous chapters; it is in this way that we can make quite precise the statement that the area of a circle can be approximated by means of polygons, as well as the theorem about the distribution of prime numbers. Of course the reader will have to take my word for this, since I cannot go into the lengthy proofs that would be involved.

In Algebra for example we determined a number in the following way: let X be the number which is such that if you divide it by 2, multiply it by 3 and add 5 to it, you get 11, i.e. let X denote the number which satisfies the equation

$$\frac{X}{2} \times 3 + 5 = 11$$

Here we have learnt another method of determining a number. The branch of Mathematics which deals with the determination of numbers with the aid of successive approximations, but at the same time with complete accuracy, is called Analysis.

Let us start on the other hand from $1\frac{1}{9}$. One whole can be divided in 9 ninths, so that

$$1\frac{1}{9} = \frac{9}{9} + \frac{1}{9} = \frac{10}{9} = 10 \div 9 = 1{\cdot}1111111 \ldots$$

indefinitely, and the identity of $1\frac{1}{9}$ and this infinitely long expansion have just received a definite meaning above.

In Mathematics this is expressed by saying that the sequence of the 'partial sums'

$$1, \quad 1{\cdot}1 = 1 + \frac{1}{10}, \quad 1{\cdot}11 = 1 + \frac{1}{10} + \frac{1}{100}, \cdots$$

* This will be referred to as 'the chocolate example' (*translator's note*).
† Approximations in general will be dealt with in the next chapter.

'tends to the limit' $1\frac{1}{9}$, or by saying that the series

$$1 + \frac{1}{10} + \frac{1}{100} + \cdots$$

is convergent and its sum is $1\frac{1}{9}$.

Here we have introduced a new kind of addition. We ought to see whether it obeys the rules of the old operations. I do not want to go into the examination of such fiddly little details; I shall just state the reply to the statement: there is no question of the old rules being obeyed. The Infinite escapes from our rules here, too, so much so that a special study has developed concerning those series of which the terms can be interchanged in any order or grouped in any way. The series we have discussed is just such a series:

$$1 + \frac{1}{10} + \frac{1}{100} + \cdots$$

But let us have a look at the series

$$1 - 1 + 1 - 1 + 1 - 1 + \ldots$$

If we change the order of the operations and group the terms together in pairs

$$\underbrace{1 - 1}_{0} + \underbrace{1 - 1}_{0} + \underbrace{1 - 1}_{0} + \ldots$$

we shall get a series consisting of zeros only, and, however many zeros we add, the result is always zero, and so the result of the addition must be zero. But if we group the terms as below

$$1 \underbrace{- 1 + 1}_{0} \underbrace{- 1 + 1}_{0} - \ldots$$

then we construct the series

$$1 + 0 + 0 + 0 + \ldots$$

and the sum of this series is obviously 1. So we cannot expect to be able to carry out operations in any order.

Some things survive, however: for example we can multiply an infinite series term by term by a number.

Let us go on playing with our result. If from

$$1 \cdot 1111111 \ldots = 1\frac{1}{9}$$

we take away the 1, we get

$$0 \cdot 1111111 \ldots = \frac{1}{9}$$

and multiplying by 9 we have

$$0 \cdot 9999999 \ldots = \tfrac{9}{9} = 1$$

Dividing by 10 (if we carry the decimal point one place to the left, imagining this decimal point following the 1 whole number, there will be 0 whole numbers) we have

$$0 \cdot 0999 \ldots = 0 \cdot 1$$

Dividing by another 10 we have

$$0 \cdot 00999 \ldots = 0 \cdot 01$$

and so on. So that the finite decimals 1, 0·1, 0·01, . . . can be written down also as infinite decimals in which after the zeros there are nothing but 9's. From this it follows straight away that every finite decimal can be written down as an infinite decimal in two ways. Let 0·2 be the finite decimal in question. We can write it first of all like this:

$$0 \cdot 2000000 \ldots$$

since adding 0 hundredths, 0 thousandths, 0 tenthousandths cannot alter the number. Another way of writing it would be

$$0 \cdot 199999 \ldots$$

since the 1 tenth, i.e. 0·1, which has been left out of 0·2 is the same as 0·099999 . . ., which has been added (it would be possible to prove that this is the only possible ambiguity that could arise in the decimal expansions of numbers).

In the series we have examined, i.e. in

$$1 + \frac{1}{10} + \frac{1}{100} + \frac{1}{1000} + \ldots$$

every term is one-tenth of the previous one, i.e. is obtained from the previous one by multiplying it by $\frac{1}{10}$. The reader should remind himself of the arithmetical series in which any two neighbouring terms have the same difference. The kind of series in which the quotient of any two neighbouring terms is the same is called a geometrical series.

We must not get conceited and think that we can now sum all infinite series. Let us have a look at the following geometrical series, for example:

$$1 + 10 + 100 + 1000 + \ldots$$

in which the quotient of neighbouring terms is 10. It is

obvious that its partial sums will eventually get larger than any individual number (for example, already from the fourth one onwards they are greater than 1000), and so this series tends to infinity. What is more, even that geometrical series in which every term is followed by that term multiplied by 1 does likewise, since in

$$1 + 1 + 1 + 1 + 1 + \ldots$$

every partial sum is greater than 1000 from the thousandth on, greater than 1,000,000 from the millionth on and so on.

If every term is followed by that term multiplied by -1, then since $1 \times (-1) = -1$, $(-1) \times (-1) = +1$, $(+1) \times (-1) = -1$ and so on, the series will be

$$1 - 1 + 1 - 1 + 1 - 1 + \ldots$$

and we already know all sorts of dreadful things about this one. Its partial sums in order are :

$$1,$$
$$1 - 1 = 0$$
$$1 - 1 + 1 = 0 + 1 = 1$$
$$1 - 1 + 1 - 1 = 0 + 0 = 0$$

and so on. We can see that these are alternately equal to 1 and to 0.

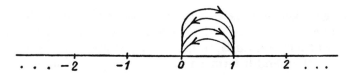

They jump about between 0 and 1 (using a longer word, they 'oscillate'), and so they do not approximate to any kind of number at all. You can have bigger jumps, even increasingly bigger jumps, if the quotient of two neighbouring terms is a negative number whose absolute value is greater than 1. The picture for this would be:

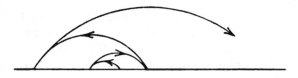

Among the series we have examined up to now there has so far been only one that we have been able to sum, i.e.

$$1 + \frac{1}{10} + \frac{1}{100} + \frac{1}{1000} + \cdots$$

This is probably to do with the fact that the terms of this series get smaller and smaller; moreover they will become as small as we please, provided we go far enough in the series. With the precision of the chocolate example, they tend to zero. (It can be shown that if something is equal to 1 as a first approximation, equal to $\frac{1}{10}$ as a second approximation, equal to $\frac{1}{100}$ as a third approximation and so on, then with perfect accuracy this thing can only be zero. I shall not formulate this always in such a lengthy way, but I shall refer to the precise formulation of the chocolate example.) In this way we can imagine that even if we have to add an infinite number of numbers, of which the terms get smaller and smaller, or rather more and more negligible, then these will influence the result less and less, and so the partial sums which include further terms of the series represent the sum of the series better and better.

But this is not really enough to make it possible to sum a series. The sequence

$$1, \ \frac{1}{2}, \ \frac{1}{3}, \ \frac{1}{4}, \ \frac{1}{5}, \cdots$$

converges likewise to zero, although more slowly than the previous one. In our first series every term is less than $\frac{1}{1000}$ from the fourth term onwards; in the series just given this is only so from the thousandth term onwards. The partial sums of the series

$$1 + \frac{1}{2} + \frac{1}{3} + \frac{1}{4} + \frac{1}{5} + \frac{1}{6} + \frac{1}{7} + \frac{1}{8} + \frac{1}{9} + \frac{1}{10} + \frac{1}{11} + \frac{1}{12} +$$

$$+ \frac{1}{13} + \frac{1}{14} + \frac{1}{15} + \frac{1}{16} + \cdots$$

nevertheless tend to infinity.

We can understand this in the following way: we know that the value of a fraction becomes less if we make the denominator bigger (if we cut the cake into more slices, the slices will be smaller). Accordingly the partial sums will be made smaller if instead of $\frac{1}{3}$ we put a smaller number, $\frac{1}{4}$, instead of each of the fractions $\frac{1}{5}$, $\frac{1}{6}$, $\frac{1}{7}$ we put $\frac{1}{8}$, instead of each of $\frac{1}{9}$, $\frac{1}{10}$, $\frac{1}{11}$, $\frac{1}{12}$ $\frac{1}{13}$, $\frac{1}{14}$, $\frac{1}{15}$, we put $\frac{1}{16}$ and so on, for in general we always go as far as the terms in whose denominators a power of 2 is found $(4 = 2^2, 8 = 2^3, 16 = 2^4)$, and replace the previous terms by this term. So the partial sums of the following series are sure to be smaller than the partial sums of our series:

$$1 + \frac{1}{2} + \frac{1}{4} + \frac{1}{4} + \frac{1}{8} + \frac{1}{8} + \frac{1}{8} + \frac{1}{8} +$$

$$+ \frac{1}{16} + \frac{1}{16} + \frac{1}{16} + \frac{1}{16} + \frac{1}{16} + \frac{1}{16} + \frac{1}{16} + \frac{1}{16} + \cdots$$

Here are the values of the groups one after the other:

$$\frac{1}{4} + \frac{1}{4} = \frac{2}{4} \qquad \text{simplified} = \tfrac{1}{2}$$

$$\frac{1}{8} + \frac{1}{8} + \frac{1}{8} + \frac{1}{8} = \frac{4}{8} \qquad \text{,,} \qquad = \tfrac{1}{2}$$

$$\frac{1}{16} + \frac{1}{16} + \frac{1}{16} + \frac{1}{16} + \frac{1}{16} + \frac{1}{16} + \frac{1}{16} + \frac{1}{16} = \frac{8}{16} = \frac{1}{2}$$

and so on. We see that every group yields $\tfrac{1}{2}$. Thus $2000 \times \tfrac{1}{2}$ is 1000; $2{,}000{,}000 \times \tfrac{1}{2} = 1{,}000{,}000$, so that the partial sums of this series, if they are long enough, become larger than any number. Even more so therefore would this be true for the larger sums of the original series.

Thus in order that an infinite series be summable it is not enough that the terms tend to zero in a half-hearted sort of way; they must approach zero at a fair rate.

12. *The line is filled up*

THE expansions of fractions into decimal form turned out to be surprisingly regular; they resulted either in finite or in recurring decimals. In the meantime we have got used to the idea of treating an infinite expansion as a single definite number, since for example starting with 1·111111 . . . we found out that this was exactly equal to 1⅑. It is almost impossible to avoid wondering whether we could imagine decimals which are not recurring; will there be no number corresponding to such an expansion?

As a matter of fact we can construct an expansion whose decimal places are quite regular, so that anyone could go on writing them down; in this way we have a conception of the whole expansion, yet we cannot find any recurring groups in it. For example

0·101001000100001000001 . . . is such an expansion.

The rule is very simple. There is always one more 0 after each successive 1; so there can be no question of recurrence, since then sooner or later the 1's occurring between the 0's would follow one another at regular intervals. This cannot be the expansion of any fraction; its partial sums cannot converge to any rational number.

I shall show that they nevertheless converge to some sort of gap in the set of rational numbers. This will show that in spite of the limitless density of the rational numbers, some gaps nevertheless occur among them.

If we stop at the tenths, we neglect a lot of places. On the other hand every partial sum is less than 0·2, since the 1 tenth would have to be followed by nothing but 9's,

0·199999999 . . .

in order for it to be equal to 0·2, according to what we have said in the previous chapter. Accordingly all further partial sums will lie between 0·1 and 0·2, that is they will be represented on the line below somewhere along its thick part

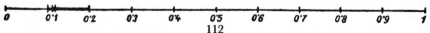

As a first approximation we could take any arbitrary point in this interval.

In the same way we can see that if we stop at the thousandths, the partial sums will be squeezed between 0·101 and 0·102. This can be shown only roughly on the figure as the points in question come so close together (their difference is one thousandth). So the points of this interval give a much better approximation; the new interval lies entirely inside the first interval. Continuing the argument, the longer and longer partial sums will be boxed in the narrower and narrower intervals:

between 0·101001   and 0·101002
between 0·1010010001 and 0·1010010002 . . .

If these intervals did not get so small so quickly, the picture would look something like this:

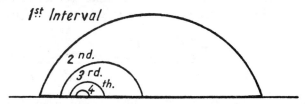

The lengths of our intervals are as follows

0·1
0·001
0·000001
0·0000000001

i.e. one-tenth, one-thousandth, one-millionth of a unit and so on. These do, of course, converge to zero (and with such terrific speed that it is impossible to follow it either by drawing or even by means of words). Our longer partial sums will therefore all crowd into these intervals, which become smaller and smaller indefinitely.

This is the same kind of boxing as little children's boxes that fit into each other. Or it is like the parcel passed round in 'pass the parcel', when each time one wrapping is taken off another wrapping is found and which you undo more and more excitedly, and yet with the accumulation of more and more wrappings. The big parcel gets smaller and smaller, but eventually one gets to the end, and there is usually some little thing inside all the wrappings, if only a little paper ball.   But you cannot go on indefinitely annoying someone with a parcel like that.

The second interval is entirely inside the first, the third one is inside the first and the second, the fourth interval is inside all previous intervals and so on.   Our intuition tells us that if we continue to construct these intervals contained inside one another, which grow smaller and smaller indefinitely, the little bit which they become in the end should be a common part of them all.   It can be proved that more than one point could not be contained in all of them.   Let us suppose in fact that I have found such a common point, and somebody comes and says that he has found a different one, a point which is different from mine, and yet is contained in all the intervals.   Of course his point would not be very far from my point.   In the figure below I must draw them at quite a distance so that we can see them easily, but the argument applies to any small distances.

However close the points are, if they are different, there will still be a certain distance between them, let us say 2 thousandths of a unit.   Let us take half of this, i.e. one-thousandth. The lengths of the intervals that are boxed inside one another tend to zero, so that sooner or later their lengths will be less than one-thousandth of a unit.   My point will be included in one of them, and even if it happens to be near the left end of an interval whose length is less than one-thousandth, this interval could never extend as far as the point lying at a distance of 2 thousandths.

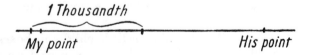

The other point therefore which has been given as a counter-example will certainly be left out of such intervals or of any smaller intervals.   So it is quite impossible that this other point is also a common point of all the intervals.

All our intervals have just one single definite point in common, and, since whichever interval we choose, the partial sums of 0·1010010001 . . . will all be in it provided we go far enough, the points corresponding to these partial sums get nearer and nearer to our point, in other words they converge to our point.

In this way we have found a point on our line of numbers to which up till now no number has corresponded. However thickly the fractions cover the line, no fraction has found its way to this point. The decimal forms of fractions are recurring, yet our 0·101001000100001 . . . which converges to this point, never becomes recurring. And yet this is a quite definite point, and is at a certain determinate distance from 0. But if we try to measure this distance, it is impossible to do so in whole units, and impossible in fractions of a whole unit. So up till now this distance has not even possessed a measure. In order to make up for the lack of this, we shall say that the measure of this distance is the 'irrational' number

$$0·101001000100001000001 . . .$$

and so we introduce this so far nameless but quite determinate something to which the rational values

$$0·1, \quad 0·101, \quad 0·101001, . . .$$

are better and better approximations. This is no less useful for the practical man or for the physicist than the expansion $\frac{4}{9} = 0·444444 . . .$ There is no degree of accuracy which could not be achieved in the approximations to this number, and we know all its digits that we are ever likely to need, since we have a picture of the pattern of the digits as a whole.

We can show in just the same way that a definite point corresponds to any infinite decimal which is non-recurring but given by some rule, i.e. such a point is at a definite distance from 0 along the line. We shall regard all such infinite decimal expansions as the measures of the corresponding distances and we shall call them irrational numbers.

Perhaps these considerations appear very abstract. Yet I once had a pupil, Eva, in my fourth form, who found out for herself that there were distances whose lengths could not be expressed either in whole units or in fractions. She was doing the following amusing puzzle. At every corner of a square fishpond there is a tree.   The problem is to make the fishpond twice as big, though it has to remain a square and the trees have to remain where they are:

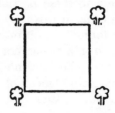

Eva found that the solution was as follows:

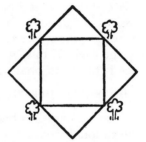

The big square is in fact twice as big as the small square, because if we draw the diagonals in the small square,

we can readily see that if we fold the four triangles so obtained outwards,

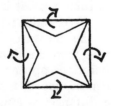

we obtain the larger square.   It is obvious that we have added to the small square as much again as its original area.

But Eva was not satisfied with this. She was curious to know how long the sides of the new fishpond would be if the old fishpond had sides whose lengths were 1 mile. The area of the old pond in this case was $1 \times 1 = 1$ square mile and so the area of the enlarged fishpond was twice as much, i.e. 2 square miles. The problem was to find a number which, squared, gave 2. This is how we arrived at the inversion of the operation of raising to a power, i.e. the extraction of roots. The problem was the calculation of $\sqrt{2}$, the number whose square is 2. If there is such a number it is denoted by $\sqrt{2}$.

So Eva started to try this and that. The side of the small square was 1 mile, the side of the larger square was obviously longer. But it could not be 2 miles long, since $2 \times 2 = 4$ and so its area would be 4 square miles. So the length of the side must be between 1 mile and 2 miles.

Then Eva tried to take a few tenths more than 1. During these trials she found that

$$1 \cdot 4^2 = 1 \cdot 4 \times 1 \cdot 4 = 1 \cdot 96 \quad \text{and} \quad 1 \cdot 5^2 = 1 \cdot 5 \times 1 \cdot 5 = 2 \cdot 25$$

$1 \cdot 96$ is less, while $2 \cdot 25$ is more than the area of the pond, which is 2 units. So the length of the side must lie between $1 \cdot 4$ and $1 \cdot 5$ miles. She then went on dividing this interval into hundredths, and it became apparent in much the same way that the length of the side must lie between

$$1 \cdot 41 \quad \text{and} \quad 1 \cdot 42$$

Continuing this for some time, Eva became more and more convinced that she would never find the number whose square was 2. 'But there must be such a number! Here it is quite clearly as the side of the larger square. I have constructed it myself!' said Eva.

Eva was right in her intuition. There is no rational number whose square is 2. Eva had proved that there was no such whole number, when she showed that the number required must lie between 1 and 2, for between 1 and 2 there are no more whole numbers. So it only remains to examine the fractions that lie between 1 and 2.

Let us simplify these until they cannot be simplified any more. Their denominator cannot in this way become 1, since, for example, $\frac{3}{1}$ is really 3 wholes, and there are no whole numbers between 1 and 2. And it is equally certain that we cannot simplify their squares, for example,

$$\left(\frac{15}{14}\right)^2 = \frac{15 \times 15}{14 \times 14}$$

and we cannot simplify for $\frac{15}{14}$, because $15 = 3 \times 5$ and $14 = 2 \times 7$, and 15 and 14 have no common prime factor. On the other hand they cannot acquire common prime factors through multiplications by themselves,

$$\left(\frac{3 \times 5}{2 \times 7}\right)^2 = \frac{3 \times 5 \times 3 \times 5}{2 \times 7 \times 2 \times 7}$$

and so there can be no question of any simplification. But a fraction which cannot be further simplified and whose denominator is not 1 cannot possibly be equal to 2.

In spite of this, however, Eva's trials are the beginnings of successive boxings, and, at the same time, the beginning of the decimal expansion of $\sqrt{2}$. The decimal form of any number between 1 and 2 is certainly going to begin like this:

$$1\cdot \ldots$$

Any number with such a beginning can be considered as a first approximation for $\sqrt{2}$.

If we know, further, that this number lies between 1·4 and 1·5, then the decimal expansion will continue like this:

$$1\cdot4 \ldots$$

and numbers with such a beginning already give a better approximation. From the fact that the number we are looking for lies between 1·41 and 1·42, we continue

$$1\cdot41 \ldots$$

Now we must divide the interval between 1·41 and 1·42 into thousandths and see which one among

1·410, 1·411, 1·412, 1·413, 1·414, 1·415, 1·416, 1·417, 1·418, 1·419

is the number whose square is smaller than 2, but such that the square of the number after it is greater than 2. These two numbers give us a box whose length is only one-thousandth, in which to put $\sqrt{2}$, and at the same time we have found the third decimal place in the expansion of $\sqrt{2}$.

There is a more mechanical method for determining the decimal expansion of $\sqrt{2}$, but the essential point can be seen through this squeezing into smaller and smaller boxes.

This process can be continued indefinitely and it gives continually better and better approximations. We know that it can never come to an end or become recurring, since $\sqrt{2}$ cannot be a rational number. Still, there it stands before us quite exactly and tangibly; we know just how big this number is that we obtained by more and more exact approximations. It is just the length of the enlarged fishpond.

The well-known theorem due to Pythagoras also helps to clarify for us what this $\sqrt{2}$ really is. Let us draw a right-angled triangle so that the sides adjacent to the right angle are both one unit, and let us draw a square on each of its sides.

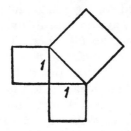

If we put in one diagonal in each of the small squares and both diagonals in the large square, we get all congruent triangles

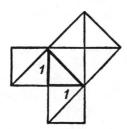

and four of these will be found in the two little squares together and four in the large square. Therefore the sum of the areas of the little squares will be the same as the area of the larger square, and since we can calculate the area of a square by squaring the length of a side, the sum of the squares of the two sides adjacent to the right angle is equal to the square of the hypotenuse (this is not only true of this special triangle but of all right-angled triangles, the proof in the general case being a little more complicated). Here the sum of the squares of the two sides adjacent to the right angle is

$$1^2 + 1^2 = 1 + 1 = 2$$

and this is just the same as the square of the hypotenuse, so that the length of the hypotenuse is $\sqrt{2}$ units.

It can be shown that operations with irrational numbers can be carried out by working on the approximate values. The approximate values are rational numbers, and these still obey the old rules of manipulation. Here we are dealing with a case in which the old rules hold even for an infinite series.

Now we can go back to the problem, left in abeyance, of whether you can always express the length of the edge of a cube or the length of the side of a rectangle in inches. The answer is that you cannot always do this, in the sense that there are distances which cannot be measured even by means of any fraction of an inch. For example if $\frac{1}{20}$th of an inch covers a certain distance 31 times, then this distance is $\frac{31}{20}$ of an inch; yet we have just seen that, if each side adjacent to the rectangle in a right-angled triangle is 1 inch, then the length of the hypotenuse cannot be expressed in terms of this unit by any rational number whatsoever. (The reason we must emphasize that it is in terms of *this* unit is because there is of course a certain length corresponding to $\sqrt{2}$ and we could choose this length as the unit, and in terms of this unit it would itself be of course expressible.)

In spite of these difficulties, we can prove with the aid of the rational approximate values with 'chocolate' precision that the old results about areas and volumes still remain valid.

I am still in the reader's debt in connexion with quadratic equations. We got into difficulty with the equation

$$(X + 3)^2 = 2$$

Now we can solve this equation. Since we now have negative numbers as well, and we know that the squares of positive and of negative numbers are positive, both $+\sqrt{2}$ and $-\sqrt{2}$ can be regarded as the number whose square is 2. Accordingly

either $X + 3 = +\sqrt{2}$ or $X + 3 = -\sqrt{2}$

and, if we take the 3 over to the right as a subtraction, we get two answers:

$$X = +\sqrt{2} - 3 \text{ or } X = -\sqrt{2} - 3$$

But the negative numbers introduce a new complication. We do not know what to do with the equation

$$X^2 = -9$$

since both $+3$ and -3 when squared give $+9$. We do not know any number whose square is -9. We shall return to this question later.

We introduced irrational numbers because we found gaps in the line of our numbers, points to which no number corresponded. The rational and the irrational numbers together (the so-called real numbers; we shall have to do with numbers whose reality is much more questionable) already fill up the line completely, since if we take any point on the line, this will be in turn between certain wholes, between certain tenths, between certain hundredths, just as was $\sqrt{2}$ that my pupil Eva examined, and these intervals give us one after the other the digits in the decimal expansion of some number. If this expansion ends somewhere (that is, if the point coincides with one of the tenths, hundredths, thousandths, . . .) or becomes recurring, then the number corresponding to our point is rational; if not, then it is irrational.

If for example we tried to shut up the point corresponding to the number $1\frac{1}{9}$ occurring in our chocolate example* in such boxes, we should see that it would come between 1 and 2, then between 1·1 and 1·2, then between 1·11 and 1·12, then between 1·111 and 1·112, and so each of

$$1, \quad 1\cdot1, \quad 1\cdot11, \quad 1\cdot111, \ldots$$

would one after the other fall into our shrinking boxes (coming just at the left-hand ends of the boxes each time). This is the background situation to the fact that these numbers give better and better approximations for $1\frac{1}{9}$, i.e. they get as near as we like to this number. Of course they yield a recurring decimal, since $1\frac{1}{9}$ is rational.

Are there a lot of irrational numbers? Even though we have so far only come across them accidentally, there must be a lot

* See pages 105–106.

of them, since we instinctively feel that if a decimal expansion turns out to be recurring, it is just chance. But we have been led astray before by such feelings; we thought it went without saying that there were more rational numbers than natural numbers, but we found eventually that all rational numbers could be arranged in a single sequence, and so could be paired off with the natural numbers: the first number in the sequence with 1, the second number with 2, and so on. We might wonder what this pairing process has to say about irrational numbers.

Let us first of all examine the rational and irrational numbers jointly, i.e. the real numbers, and let us imagine them all in their decimal expansions. Of these let us restrict ourselves to numbers between 0 and 1, i.e. to numbers which begin with 0 whole-number parts, so that we do not need to bother with the whole parts. I am suggesting that even this section of real numbers is more numerous than all the natural numbers, i.e. that you cannot arrange them in a sequence without leaving some real numbers out of it.

Let us suppose that somebody tells me that I am wrong, that this somebody knows of an example to the contrary. He claims to have constructed a sequence of all the real numbers (which have 0 whole parts) without leaving any of them out. He writes down this sequence, by giving a few numbers from which a definite regularity can be seen which enables anyone to continue the sequence as far as he likes. But he can give in this way the individual numbers too, since these are already infinite decimals. Let us say the sequence begins as follows:

> the first number: 0·1
> the second number: 0·202020 . . .
> the third number: 0·3113111311113 . . .
>
> .   .   .   .

and these are supposed to be continued according to some kind of rule so that sooner or later every real number will be included in the sequence.

Whatever this rule may be, we can immediately construct a real number with 0 whole part which has certainly been left out of the sequence.

First of all we complete the finite decimals by joining a lot of innocent 0's indefinitely, in this way:

the first number: 0·100000000000000000 . . .
the second number: 0·202020202020202020 . . .
the third number: 0·311311131111311111 . . .

. . . .

Now we can begin: the first figure of our number is:

0· . . .

What shall we write in the tenths place? We can have a look and see what number appears at the tenths place in the first number of the sequence of the example, and we, to the contrary, can write *something else*, only we must be careful never to write a 0 or a 9.   In order to be more definite, since the first digit in the counterexample is 1, let us write a 2 in the tenths place (I could here or in any of the others have written any of the numbers 3, 4, 5, 6, 7 or 8).   If there had been any other number in the tenths place in the counterexample, we should have chosen 1 for our tenth.   So our number so far is

0·2 . . .

We can fill in the hundredths place by looking at the hundredths place in the second number of the counterexample, and again we can write *something else* in the hundredths place of our number.   Let us keep to 1 and 2.   At the hundredths place in the second number of the counterexample here quoted there is a 0; since this is not a 1, we write a 1 in its place (if there had been a 1 there, we should have written a 2).   Our number then is continued like this:

0·21 . . .

We can continue with this indefinitely: we shall write a 2 in the thousandths place, since there is 1 thousandth in the third number, so that up to three places of decimals our number is

0·212 . . .

and now anyone can continue the process as far as he likes: if the numbers in the counterexample succeed one another according to some sensible rule, then we cannot get stuck in the construction of our number.   In this way we obtain an infinite decimal with 0 whole-number part which certainly has been left out of the sequence.   Our number differs from the first number at least in the tenths place, from the second number at least in the hundredths place, from the third at least in the thousandths place, so it differs from every one of them in at least one digit.   It is even impossible for our number to differ in form only and not in value from any of the numbers in the counterexample, since such ambiguity can occur only with numbers whose expansions consist of either all 0's or all 9's from a certain place onwards, and our number consists of only 1's and 2's.

So however anyone were to try to pair off the real numbers with the natural numbers 1, 2, 3, 4, 5, . . . by writing them in a sequence, there would always be a real number that was left out of it.   The real numbers are more numerous than the natural numbers.   If we do not restrict ourselves to those whose whole-number parts are 0, this must of course be even more so.

We have actually proved this only about rational and irrational numbers taken together.   But we already know that the rational numbers are countable, i.e. they can be written out in the form of a sequence.   If the irrational numbers could also be written out in the form of a sequence, then it would be very easy to unite these two sequences by taking numbers alternately from each to make the new sequence (for example we can unite the sequences of the positive and negative whole numbers

$$1, \quad 2, \quad 3, \quad 4, \quad 5, \ldots$$

and $$-1, \quad -2, \quad -3, \quad -4, \quad -5, \ldots$$

into a single sequence by writing:

$$1, \quad -1, \quad 2, \quad -2, \quad 3, \quad -3, \quad 4, \quad -4, \quad 5, \quad -5, \ldots)$$

The joint sequence of rational and irrational numbers would contain all real numbers, but we have just proved that it is impossible to have one sequence to do this.   Therefore the set of irrational numbers cannot be written out in a sequence even on their own; they cannot be countable and so they are more numerous than the rational numbers.

When we introduced irrational numbers it was therefore not just a question of filling in a few gaps in the everywhere dense set of rational numbers.   The irrational numbers spread continuously over the whole line in spite of the density of the rational numbers, so now the rational numbers are the raisins scattered in the cake of irrational numbers.   This seems a little similar to the old hypothesis about ether, which was that ether takes up all the room in the atmosphere without leaving any gaps, and yet the apparently ubiquitous molecules of air are still swimming dispersed in it.

13. *The charts get smoothed out*

In going through all the items representing my debt to the reader, I suddenly thought of the poor lonely 1 at the top of the Pascal Triangle:

$$1$$
$$1 \qquad 1$$
$$1 \qquad 2 \qquad 1$$
$$1 \qquad 3 \qquad 3 \qquad 1$$

$$\cdot \quad \cdot \quad \cdot \quad \cdot \quad \cdot \quad \cdot \quad \cdot$$

We proved that, starting from the second row, the sum of the terms in each row is successively

$$2^1, \quad 2^2, \quad 2^3, \ldots$$

and if we wanted to fit the topmost 1 into this order then its value would have to be 2^0.   But 2^0 has so far had no meaning; you cannot multiply something by itself 0 times, and up till now we have not felt the necessity of giving such an expression a meaning.

Let us devote a little more time to the operation of raising to a power. We remember how easy it was to multiply the powers of a certain number by each other; we just had to add the exponents. For example:

$$3^2 \times 3^4 = \underbrace{3 \times 3} \times \underbrace{3 \times 3 \times 3 \times 3} = 3^6$$

and $$6 = 2 + 4$$

Other operations too can be easily carried out, if we are dealing only with the powers of a single number.   For example:

$$\frac{3^6}{3^2} = \frac{3 \times 3 \times 3 \times 3 \times 3 \times 3}{3 \times 3}$$

and if we cancel 3×3, we obtain

$$\frac{3 \times 3 \times 3 \times 3}{1} = 3 \times 3 \times 3 \times 3 = 3^4$$

so that $$\frac{3^6}{3^2} = 3^4 \quad \text{and} \quad 4 = 6 - 2$$

and the division can be carried out by subtracting the exponents.

Or $(3^2)^4 = 3^2 \times 3^2 \times 3^2 \times 3^2$

$$= \underbrace{3 \times 3}_{} \times \underbrace{3 \times 3}_{} \times \underbrace{3 \times 3}_{} \times \underbrace{3 \times 3}_{}$$

$$= 3^8$$

and $8 = 2 \times 4$.

If, therefore, we have to raise a power to another power, we simply multiply the exponents. For this reason, it is worth while making a table of the powers of a certain base. Let us choose 2 as our base; we can easily calculate its successive powers:

$2^1 =$	2
$2^2 =$	4
$2^3 =$	8
$2^4 =$	16
$2^5 =$	32
$2^6 =$	64
$2^7 =$	128
$2^8 =$	256
$2^9 =$	512
$2^{10} =$	1024
$2^{11} =$	2048
$2^{12} =$	4096

If we have to multiply two numbers, and we are lucky, we can extract the result from this table without any trouble. For example if we have to multiply

$$64 \times 32$$

then we are in luck since both numbers occur in our table. The corresponding exponents are 6 and 5, and it is no great difficulty to add these, giving 11. One look at the 11th row gives the result

$$2048$$

Or if on the other hand we need to square 32, the corresponding exponent is 5. We can multiply this by 2 in a fraction of a second; we get 10, and from the 10th row we can read off the result:

.   .   .   . $32^2 = 1024$

This is really child's play; the pity of it is that all numbers do not figure in the table. It might be worth while to extend the meaning of raising to a power so that every number (for example even 3) should be expressible as a power of 2.

In this way we arrive at a different inversion of the operation of raising to a power. We are now looking for the exponent, to which we should need to raise 2 in order to get 3 as a result.

This operation is called taking the logarithm, and the result of the operation is the logarithm.

The most inconvenient things from the point of view of calculations are fractions. These have not yet occurred in our table. The smallest power of 2, 2^1 is equal to 2 wholes. In the attempt to find new meanings our leading consideration will be, as before, that numbers greater than 2 should be expressible as greater powers of 2, so that we shall not have to look for them all over the table. We must therefore introduce powers of 2 which are less than 1 if we want to express fractions as well. If we go backwards in whole steps, the symbols

$$2^0, \quad 2^{-1}, \quad 2^{-2}, \quad 2^{-3}, \ldots$$

are patiently queuing up to be given meanings.

In this extension of operations we must be extra careful to see that the old rules remain valid; we must not lose sight of our aim: we want calculations with the new powers to be just as convenient as they were with the old ones.

Among other things we must be careful that if we multiply any power of 2 by 2^0, we get the same answer as if we added 0 to the exponent. But the adding of 0 does not alter anything, so that we need to give 2^0 that meaning which will ensure that if we multiply by it, the value of the number multiplied is not altered. The multiplier which does not change the value of a number is of course 1, and thus we must define 2^0 (and similarly the 0th power of any other base) by the requirement

$$2^0 = 1$$

With this definition the Pascal Triangle acquires a uniform meaning.

When we want to give a meaning to 2^{-1}, then we must be careful to ensure that

$$2^1 \times 2^{-1} = 2^{1+(-1)} = 2^0 = 1$$

If, on the other hand, we take 2^1 over to the other side in the equation

$$2^1 \times 2^{-1} = 1$$

the equation becomes

$$2^{-1} = \frac{1}{2^1}$$

Similarly from the requirement

$$2^2 \times 2^{-2} = 2^{2+(-2)} = 2^0 = 1$$

we get
$$2^{-2} = \frac{1}{2^2}$$

and from the requirement

$$2^3 \times 2^{-3} = 2^{3+(-3)} = 2^0 = 1$$

we get
$$2^{-3} = \frac{1}{2^3}$$

and so on. If we wish to preserve intact all our convenient processes, we must interpret the powers with negative exponents as a division of 1 by the corresponding power with positive exponents. In this way our table is extended backwards as well, and includes some fractions:

$$2^{-3} = \frac{1}{2^3} = \frac{1}{8} = 1 \div 8 = 0\cdot125$$

$$2^{-2} = \frac{1}{2^2} = \frac{1}{4} = 1 \div 4 = 0\cdot25$$

$$2^{-1} = \frac{1}{2^1} = \frac{1}{2} = 1 \div 2 = 0\cdot5$$

$$
\begin{aligned}
2^0 &= & 1 \\
2^1 &= & 2 \\
2^2 &= & 4
\end{aligned}
$$

. . . .

This is quite a help for reckoning with the fractions $\frac{1}{2}$, $\frac{1}{4}$, $\frac{1}{8}$, . . . i.e. with the decimals $0\cdot5$, $0\cdot25$, $0\cdot125$, . . .

But there are still big gaps between the numbers in our table, for example $2^1 = 2$, $2^2 = 4$. If we want to write a number lying between 2 and 4 as a power of 2 (for example 3, or $2\cdot7$), then this is only possible, following the previous pattern, by using a power somewhere between 1 and 2. For example the number $1\frac{1}{2}$ lies between these two numbers, and since $\frac{2}{2} = 1$, this is equal to $\frac{3}{2}$; in this manner we must interpret the $\frac{3}{2}$th power of 2, and in general all fractional powers.

The interpretation will be decided by the consideration that we are still anxious to preserve the rule of raising a power to a power. If this is to remain valid, then

$$\left(2^{\frac{3}{2}}\right)^2 = 2^{2 \times \frac{3}{2}} = 2^{\frac{6}{2}} = 2^3$$

so that $2^{\frac{3}{2}}$ can be the only number whose square is 2^3, but this is the number which we denoted by $\sqrt{2^3}$ and so

$$2^{\frac{3}{2}} = \sqrt{2^3} = \sqrt{8}$$

and calculating this $\sqrt{8}$ to one place of decimals, we get 2·8. However, since

$$\frac{3}{2} = 3 \div 2 = 1\cdot 5$$

(we shall have to do our calculations with exponents, and it is easier to handle decimals than fractions like $\frac{3}{2}$), we can insert a new row between the rows for 2^1 and 2^2 as follows:

$$
\begin{aligned}
2^1 &= 2 \\
2^{1\cdot 5} &= 2\cdot 8 \\
2^2 &= 4
\end{aligned}
$$

Our secret ambition to write 3 as a power in this way has not been achieved, although 2·8 is quite near it. It can be shown that you cannot write down 3 as any kind of fractional power of 2 exactly, but it can be approximated to any desired extent by such fractional powers. We define the power to an irrational exponent by means of such approximations.

This is the fundamental idea behind the preparation of logarithm tables. The old logarithm tables were in fact prepared in just this way. The tables known in secondary schools all have the base 10 (the base is not even indicated, only the exponent). Here the games with our fingers have entailed considerable sacrifices. There are much longer gaps between the powers of 10, i.e. between 10, 100, 1000, . . . than between the powers of 2, and it is much more trouble to fill in these gaps.

In certain logarithm tables there are logarithms to the base 'e', called natural logarithms. This number 'e' is an irrational number which begins like this: 2·71. . . . What kind of thought process leads to taking such a number as a natural base?

There are many avenues that lead to an understanding of this, but I feel that the following is the best.

10 is not a very good number for the calculation of logarithms. It might in fact be quite an idea to take a number even less than 2 as our base, then the gaps between the whole-number powers of such a base will be even smaller. We cannot of course go as far as 1 itself, since all powers of 1 are 1, and it is not a very good idea to go below 1, since if we raise a proper fraction to a power, we make it smaller, for example $(\frac{1}{2}) \times (\frac{1}{2}) = \frac{1}{4}$. Let us try $1 \cdot 1$; this will be easy since we already know the powers of 11 from the Pascal Triangle. We just have to be careful about putting in the decimal point and to realize that every time we multiply by a tenth it is really a division by 10, so the decimal point will shift one place to the left every time. Let us also not forget that the 0th power of any base is 1.

$$1 \cdot 1^0 = 1$$
$$1 \cdot 1^1 = 1 \cdot 1$$
$$1 \cdot 1^2 = 1 \cdot 21$$
$$1 \cdot 1^3 = 1 \cdot 331$$
$$1 \cdot 1^4 = 1 \cdot 4641$$

$$. \quad . \quad . \quad . \quad .$$

These powers grow very slowly, and we have already a whole host of numbers between 1 and 2 before we need to start on the troublesome business of filling in the gaps.

Of course a still smaller number, even nearer to 1, would be better still. Let us try the base $1 \cdot 001$ (here we separate the elements in the Pascal Triangle by pairs of zeros):

$$1 \cdot 001^0 = 1$$
$$1 \cdot 001^1 = 1 \cdot 001$$
$$1 \cdot 001^2 = 1 \cdot 002001$$
$$1 \cdot 001^3 = 1 \cdot 0030003001$$

$$. \quad . \quad . \quad . \quad . \quad . \quad .$$

This is already a terrific density, for these powers grow at such a snail's pace that we might begin to wonder whether they will ever reach 2. But it is possible to prove that the powers of any number greater than 1, even if very little greater, tend to infinity, although extremely slowly.

This table still has a certain aesthetic drawback. Just on

account of this very slow growth, disproportionately large exponents correspond to the small numbers. We need to go about as far as the thousandth power to reach 2. If the exponents were a thousand times as big, they would behave in a more harmonious way. But this is quite easily achieved: let us raise the base to the thousandth power. Since

$$(1{\cdot}001^{1000})^{\frac{1}{1000}} = 1{\cdot}001^{1000 \times \frac{1}{1000}} = 1{\cdot}001^{\frac{1000}{1000}} = 1{\cdot}001^1$$

$$(1{\cdot}001^{1000})^{\frac{2}{1000}} = 1{\cdot}001^{1000 \times \frac{2}{1000}} = 1{\cdot}001^{\frac{2000}{1000}} = 1{\cdot}001^2$$

and so on, so the base $1{\cdot}001^{1000}$ has in fact only to be raised to one-thousandth of the power that $1{\cdot}001$ has to be raised to, in order to get the same result.

When we raise the base $1{\cdot}001^{1000}$ to powers, we can proceed in steps of one-thousandth. In decimal form

$$\frac{1}{1000} = 0{\cdot}001, \quad \frac{2}{1000} = 0{\cdot}002, \quad \frac{3}{1000} = 0{\cdot}003, \ldots$$

so that, using the connexions just established with the powers of the base $1{\cdot}001$, we have

$$
\begin{aligned}
(1{\cdot}001^{1000})^0 \quad &= 1{\cdot}001^0 = 1 \\
(1{\cdot}001^{1000})^{0{\cdot}001} &= 1{\cdot}001^1 = 1{\cdot}001 \\
(1{\cdot}001^{1000})^{0{\cdot}002} &= 1{\cdot}001^2 = 1{\cdot}002001 \\
(1{\cdot}001^{1000})^{0{\cdot}003} &= 1{\cdot}001^3 = 1{\cdot}003003001
\end{aligned}
$$

. . . .

The exponents and the corresponding numbers do not grow disproportionately, and the density has been unimpaired.

It is clear that the bases

$$1{\cdot}0001^{10000}, \quad 1{\cdot}00001^{100000}, \quad 1{\cdot}000001^{1000000}, \ldots$$

will do better and better for our purpose, and it can be proved that this sequence converges to an irrational number beginning with $2{\cdot}71. \ldots$ This number plays a very important role in Mathematics; it has received the distinction of a special name, it is called 'e'. The logarithms to this base e are called natural logarithms, as it is the search for more and more suitable bases that leads to them so naturally.

We filled in the gaps in the definition of power for the sake of logarithms, and now powers to any exponent have a meaning, not only powers to whole-number exponents. In this way we are now in a position to complete the very incomplete chart of the power function. We can deal with equations, and so we can write down this function in the form of an equation. Let the base be 2 again; we shall be varying the exponent. As this will be some unknown number, so I shall denote it by X, and the value of the power will vary, depending on this X. We shall call this value Y, i.e.

$$Y = 2^X$$

We shall represent the values of X by means of units like this ⊢————⊣ along a horizontal line (we can now put a 0 on this, as well as negative numbers to the left of 0), and the values of Y we shall represent upwards by means of units like this: ⏋

If $X = -3$ then $Y = 2^{-3} = \dfrac{1}{2^3} = \dfrac{1}{8}$

„ $X = -2$ „ $Y = 2^{-2} = \dfrac{1}{2^2} = \dfrac{1}{4}$

„ $X = -1$ „ $Y = 2^{-1} = \dfrac{1}{2^1} = \dfrac{1}{2}$

„ $X = 0$ „ $Y = 2^0 \phantom{=\dfrac{1}{2^1}} = 1$

„ $X = 1$ „ $Y = 2^1 \phantom{=\dfrac{1}{2^1}} = 2$

„ $X = 2$ „ $Y = 2^2 \phantom{=\dfrac{1}{2^1}} = 4$

„ $X = 3$ „ $Y = 2^3 \phantom{=\dfrac{1}{2^1}} = 8$

so that at the points, $-3, -2, -1, 0, 1, 2, 3,$ we must measure upwards the following number of units:

$$\frac{1}{8}, \ \frac{1}{4}, \ \frac{1}{2}, \ \ 1, \ 2, \ 4, \ 8$$

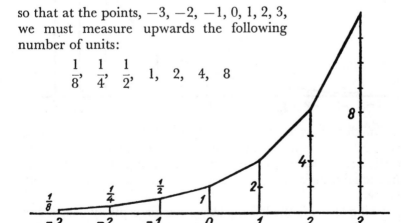

We could even choose intermediate values for X.   For example we have already seen that

$$2^{1\frac{1}{2}} = 2^{\frac{3}{2}} = \sqrt{2^3} = \sqrt{8} = 2\cdot8 \ldots$$

Similarly we can calculate values between other whole numbers.   Accurate to one place of decimals we have:

$$\begin{aligned}
&\text{If } X = -\,2\tfrac{1}{2} \text{ then } Y = 0\cdot2 \\
&\text{,, } X = -\,1\tfrac{1}{2} \text{ ,, } \quad Y = 0\cdot4 \\
&\text{,, } X = -\,\tfrac{1}{2} \text{ ,, } \quad Y = 0\cdot7 \\
&\text{,, } X = \quad \tfrac{1}{2} \text{ ,, } \quad Y = 1\cdot4 \\
&\text{,, } X = \quad 1\tfrac{1}{2} \text{ ,, } \quad Y = 2\cdot8 \\
&\text{,, } X = \quad 2\tfrac{1}{2} \text{ ,, } \quad Y = 5\cdot7
\end{aligned}$$

With these results let us now complete the graph we obtained before.   At the points $-2\tfrac{1}{2}$, $-1\tfrac{1}{2}$, $-\tfrac{1}{2}$, $\tfrac{1}{2}$, $1\tfrac{1}{2}$, $2\tfrac{1}{2}$, we must measure upwards respectively

$$0\cdot2, \quad 0\cdot4, \quad 0\cdot7, \quad 1\cdot4, 2\cdot8, 5\cdot7 \text{ units}$$

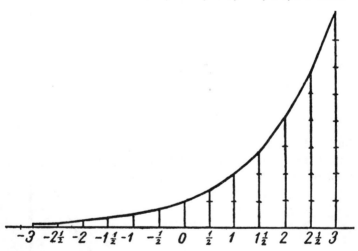

On this chart there are hardly any 'elbows' at the points where the straight segments meet.   If we continue these insertions, at least in our imagination, through all rational and irrational values of X, the chart will become a single smooth curve.

Going towards the left the curve gets perceptibly nearer and nearer to the horizontal line, but it will never reach it.   We have not found any power to which we could raise 2 and get zero, and only those points can lie on this line to which Y values of zero height correspond.

We can see the same thing happening on the chart for division which we have put aside. While we only had whole numbers this was not really possible to see. Let for example the dividend be 12 (we know that this has a lot of divisors); we are going to vary the divisor, so let us call it X. The result of the division will vary according to the divisor chosen; this will be the quotient, we shall call it Y:

$$Y = \frac{12}{X}$$

If $X = -12$, then $Y = \frac{12}{-12} = -1$

because $(-12) \times (-1) = +12$

„ $X = -6$, „ $Y = \frac{12}{-6} = -2$ for similar reasons

„ $X = -4$, „ $Y = \frac{12}{-4} = -3$

„ $X = -3$, „ $Y = \frac{12}{-3} = -4$

„ $X = -2$, „ $Y = \frac{12}{-2} = -6$

„ $X = -1$, „ $Y = \frac{12}{-1} = -12$

„ $X = 1$, „ $Y = \frac{12}{1} = 12$

„ $X = 2$, „ $Y = \frac{12}{2} = 6$

„ $X = 3$, „ $Y = \frac{12}{3} = 4$

„ $X = 4$, „ $Y = \frac{12}{4} = 3$

„ $X = 6$, „ $Y = \frac{12}{6} = 2$

„ $X = 12$, „ $Y = \frac{12}{12} = 1$

We measured the positive Y's upwards from the axis; let us draw the negative ones downwards, so that at the points

| $-12,$ | $-6,$ | $-4,$ | $-3,$ | $-2,$ | $-1,$ | we draw |
| $-1,$ | $-2,$ | $-3,$ | $-4,$ | $-6,$ | -12 | units downwards |

and at the points 1, 2, 3, 4, 6, 12   we draw
                  12, 6, 4, 3, 2, 1   units upwards.

Let the unit be ⊢⊣ in all directions

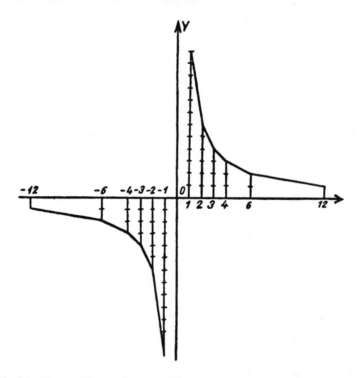

We hardly need any intermediate values; the curve is already getting nice and smooth, but it might be worth while to study its ends a little further. Here it is useful to draw another straight line upwards through the point 0. In this case we call the horizontal line the X-axis and the line perpendicular to it the Y-axis. We can see that each of the two parts of the curve gets near to both the X and the Y axes without any chance of reaching either of them; we call these lines the asymptotes of the curve. In fact, if we go farther along the X-axis towards the right, e.g. if $X = 24$,

$$Y = \frac{12}{X} = \frac{12}{24}$$

We can cancel the factor 12, giving

$$Y = \frac{1}{2}$$

If $X = 36$, and again we cancel the factor 12, we have

$$Y = \frac{12}{36} = \frac{1}{3}$$

if $X = 48$,

$$Y = \frac{12}{48} = \frac{1}{4}$$

and so on. The farther we go along the X-axis, the smaller will the Y-values become, but they will never become zero, since however we divide 12 into parts, each part will still have some size. In the same way we can go a long way in the negative direction, and we shall get the values

$$-\frac{1}{2}, \quad -\frac{1}{3}, \quad -\frac{1}{4}, \ldots$$

which tend to zero but never become zero. The other part of the curve approaches the X-axis from below more and more closely but never reaches it.

If, on the other hand, $X = \frac{1}{2}$, we know that in one whole there are 2 halves, so in 12 wholes there will be 12×2 halves, i.e. 24 halves, and thus

$$Y = 24$$

In the same way we can see that in 12 there are 36 thirds and 48 quarters, so that

$$\text{if } X = \frac{1}{3}, \text{ then } Y = 36$$

$$\text{,, } X = \frac{1}{4}, \text{ ,, } \quad Y = 48 \text{ and so on}$$

According to this if X gets nearer and nearer to 0, the corresponding Y grows taller and taller; the curve cannot reach the

Y-axis, since this could happen only if $X = 0$. But in this case we should have $Y = \frac{12}{0}$ and we should be up against the permanent and insurmountable prohibition: you shall not divide by zero!

The check for division is multiplication: $20 \div 5 = 4$ because $5 \times 4 = 20$. What are the usual things people say?

$5 \div 0 = 0$. Check: $0 \times 0 = 0$ and this is not 5.

Or $5 \div 0 = 5$. Check: $0 \times 5 = 0$ and this is not 5.

Or $5 \div 0 = 1$. Check: $0 \times 1 = 0$ and this is not 5.

Whatever number we multiply by 0, the result is always 0, so we cannot divide 5 by 0.

Let us think this over a little. If a number is very small, it goes into 5 a great number of times; the smaller the number that we divide by, the larger number we get as the result. If there were a greatest number, this would be the result of the division by the smallest number, by zero. But there is no greatest number.

But perhaps we might still be able to divide 0 itself by 0? Let us have a try: $0 \div 0 = 1$, check: $0 \times 1 = 0$; this seems to be right. But supposing I said

$$0 \div 0 = 137$$

this is also right, since 0×137 is also 0. So here we get into different sort of trouble. The result is quite indeterminate, the check gives every answer as correct. So in every case the prohibition is upheld. An amusing students' publication once formulated it in the following way: 'By every number canst thou divide, but by 0 shalt thou not divide!' said the Lord when he placed Adam in the Garden of Eden.

One might think that, since it is so strictly forbidden, it would not occur to anyone to divide by 0. So unashamedly perhaps not, but sometimes 0 turns up with a mask on, for example in the following form:

$$(x + 2)^2 - (x^2 + 4x + 4)$$

Not everybody would recognize it immediately, although here we have subtracted from $(x + 2)^2$ its own expanded form. There is always some division by some such hidden zeros in the 'proofs' where it is proved for example that $1 = 2$. In Mathematics, if we make just one mistake, if we admit just one statement which is in contradiction to the other statements, then it becomes possible to prove anything at all, even that $1 = 2$.

Let us try to remember the picture of the curve we have been studying. (I shall tell the reader its name; it is called a hyperbola.) Then we shall not forget about this prohibition. The first thing one notices about the curve is that it is in two parts. Each branch proceeds smoothly and continuously, but at the point 0 we see a terrible tear, a wound stretching to Infinity: the left-hand branch runs downwards, the right-hand branch upwards towards Infinity. And between them stands the Y-axis, like a drawn sword: 'You can approach, but you shall not come to the zero divisor!'

14. Mathematics is one

JUST because we can write down functions in the form of equations, we must not jump to the conclusion that such formulae play a decisive role in the determination of a function. Let the reader try, for example, to express the following function Y of X by means of some simple formula: every time X is a rational number, let the value of Y be 1, and every time X is an irrational number, let the value of Y be 0. (This is called a Dirichlet type of function.) The determination is unimpeachable: the value of Y depends only on what kind of an X we have chosen, and to every X corresponds quite a definite Y, for example if $X = 1 \cdot 5$, then $Y = 1$, if $X = \sqrt{2}$, then $Y = 0$. Nevertheless it is a very difficult problem to find a formula for this function, and unfortunately we cannot even represent it by a graph; it jumps about between 0 and 1 with such crazy frequency, rational as well as irrational numbers being distributed densely everywhere.

The essence of the concept of a function is the pairing of the Y values with the corresponding X values. It may happen that X cannot assume every single value; we know already from the function given by the equation $Y = \dfrac{12}{X}$ that it leaves out the value 0; the function is not defined for $X = 0$. Every time we define a function we need to state from what set of numbers X may be chosen, and instructions must be given which will make it clear what the number Y will be with which X is paired.

It is always a great help if we can draw a graph of the function. A good graph will tell us more than any detailed, verbose description.

Let us define for example the following function: whatever X may be, let Y be equal to the whole-number part of the X. For example:

$$\text{if } X = 5 \cdot 45, \text{ then } Y = 5$$
$$\text{if } X = \sqrt{2}, \text{ then } Y = 1$$

since we have already seen that $\sqrt{2} = 1 \cdot 4 \ldots$

Let us try to draw the graph of this function:

if $X = 0$ then $Y = 0$
if $X = 0.1$ then $Y = 0$
if $X = 0.9999$ then $Y = 0$

we can see that $Y = 0$ until X reaches 1, after this

if $X = 1$ then $Y = 1$
if $X = 1.001$ then $Y = 1$
if $X = 1.99$ then $Y = 1$

so that $Y = 1$ until X reaches 2, and so on; similarly in the negative direction. The graph will look like this:

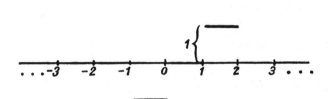

The curve consists of the above separate horizontal segments. One glance at the curve tells us everything about the function. Where the curve is broken, the value of the function jumps by one, and it remains constant along the horizontal lines. We can see that functions can have not only such infinite tears as $Y = \dfrac{12}{X}$ has at $X = 0$; they can also have more moderate ones. The graphs of both these functions do at least proceed smoothly and continuously along the untorn portions; the Dirichlet type of function, on the other hand, is not continuous anywhere. It is impossible to find an interval, however short, which would not contain rational as well as irrational points, and the value of the function is bound to jump while passing from one to the other.

We must not be led to believe that, if a function can be expressed by means of a simple formula, then by taking our points close enough to each other, its graph will be smoothed out into a curve without any 'elbows'. Suppose, for example, that we define a function in the following way: Whatever X may be, let Y be equal to the absolute value of X, that is to the value of X without regard to its algebraic sign. There is an accepted notation for the absolute value; we put a little vertical line before and after the number; for example

$$| -3 | = 3$$
$$| +3 | = 3$$

and of course $\qquad | \; 0 \; | = 0$

The function we have just defined can be expressed therefore by means of the following simple formula:

$$Y = | \, X \, |$$

Accordingly, while X runs through the values

$$-4, \quad -3, \quad -2, \quad -1, \quad 0, \quad 1, \quad 2, \quad 3, \quad 4$$

Y assumes the values

$$4, \quad 3, \quad 2, \quad 1, \quad 0, \quad 1, \quad 2, \quad 3, \quad 4$$

The graph would look like this:

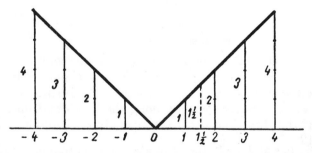

The picture of our function is therefore two straight lines leaning away from each other. This is not affected even by inserting other intermediate values. For example

if $X = 1\frac{1}{2}$, then $Y = | \, 1\frac{1}{2} \, | = 1\frac{1}{2}$

This value has been drawn in with a dotted line: the point obtained again falls right on one part of the 'elbow'.

Such geometrical representation gives a very vivid picture of the function, even if not an accurate one. Our pencils are not able to draw quite thinly enough, our rulers are not dead straight, our eyes as well as our hands are imperfect. But there are certain things that Geometry has to say about figures which have nothing to do with actual drawing. Once we get to know the geometrical properties of a hyperbola, and that the graph of $Y = \dfrac{12}{X}$ is a hyperbola, then we shall know almost everything about our function.

But even Geometry quite often looks to other branches of Mathematics for help. For example it borrows the formula when its aim is to co-ordinate the discussion of some matter; we have already seen how one formula can state a lot of different problems at one and the same time. We saw this in calculating areas and volumes. Mathematics is one, it is not split into two separate sciences, Geometry and Algebra, as children believe, particularly if the teacher has divided the syllabus in such a way that for instance there is Algebra on Mondays and Fridays and Geometry on Wednesdays; in this way Mathematics is indeed split into two subjects.

One of the bridges which joins Geometry to the other branches is the co-ordinate system: the two perpendicular straight lines passing through the point 0, the X and the Y axes, which we have already used for the description of the hyperbola. These axes give us a method of characterizing the points of the plane with the aid of numbers. We can conceive them as two paths which cross a field. If I have found a bird's nest in one of the bushes in the field, I can make a note of its position by going straight to one of the paths with as even paces as possible and counting these paces, and then counting those paces which I must take to the intersection of the paths:

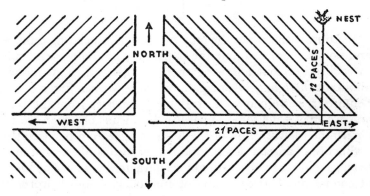

Now if I wish to direct someone else to the nest, I know that if he walks 21 paces from the crossing of the paths towards the East and then 12 paces towards the North, he will be certain to find the spot. These two directed numbers are the 'co-ordinates' of the required point. In Geometry it is usual to determine the directions by means of the signs '+' and '−', so that the positive directions point towards the right and upwards, the negative directions towards the left and downwards. Instead of paces it will be necessary to use a certain definite unit, and the co-ordinates will be measured in terms of such a unit. In this way a definite pair of numbers corresponds to every point of the plane, and one definite point to every pair of numbers. The path travelled in the direction of the *X*-axis (this one is always given first) is the *X*-co-ordinate of the point, the path travelled in the direction of the *Y*-axis is the *Y*-co-ordinate of the point.

The reader will find the co-ordinates of a number of points written by the corresponding points in the figure below; it is a good idea to get some practice in this:

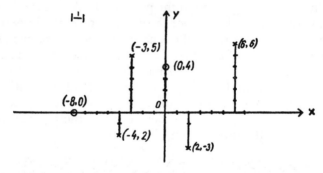

(Naturally this is not the only method of associating numbers and points. For example the paths may not be perpendicular to each other, but following them we can still find our way about; or there may be one path on which we can find a certain definite tree; we can walk straight to this tree from the bush, and we may have some apparatus by means of which we can determine the direction of the bush as seen from our tree.)

Since we can characterize points by means of pairs of numbers, we have a ready method for characterizing lines by means of connexions between pairs of numbers; in other words by

means of equations.   Consider for example the straight line
which passes through the starting point and the point (1, 1):

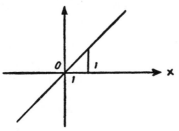

If this were a railway line, its gradient would be denoted by

$$1 : 1$$

This means that while we walk one yard along a horizontal path
next to the railway line, the line itself rises a yard.   Since this
slope rises quite evenly, after 2 yards it will also rise 2 yards,
after 3 yards it will rise 3 yards and so on.   So all the points
which lie on our straight line are characterized by the fact that
their two co-ordinates are equal.   At every such point

$$Y = X$$

Outside our line there are no points in the place whose co-
ordinates are equal.   If we join any point outside our line to
the starting point, we obtain a different slope, possibly even a
downward slope, e.g.

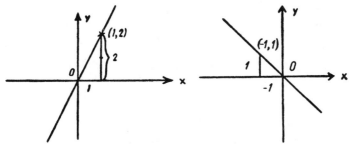

Here in the first figure the rise is in the ratio 2 : 1 all the time,
so the Y co-ordinate of any point lying on this straight line is
twice its X co-ordinate; in the second figure the slope is really
1 : 1, but the slope falls away instead of rising, and we ought
really to denote this slope by 1 : (-1) in our system of co-
ordinates; the effect of this is that the co-ordinates of any

point on this line are equal in absolute value but different in sign, and so they are not actually equal.

We see that the points outside our original line cannot have equal co-ordinates. The equation

$$Y = X$$

characterizes completely the points of our original straight line, and we are justified in saying that this is the equation of our straight line.

In the meantime we also happen to have found out the equations of the other two straight lines which we have drawn,

the one whose slope is 2 : 1 is $\quad Y = 2X$

(we shall come across this again, so please try to recognize it then) and

the one whose slope is 1 : (-1) is $\quad Y = -X$

Let us shift the straight line whose slope is 2 : 1 up a little, say by three units, but in such a way as not to change its direction.

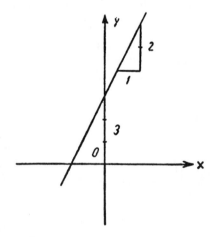

Its slope has not altered, as we can readily verify by starting from any of its points and going one unit to the right, and noticing that during this time the line has risen two units. The only thing that makes this different from the previous situation is that every point has been pushed into a position 3 units higher than it was before, and so the Y co-ordinate of every point has been increased by 3. The Y that was $2X$ before has now become $2X + 3$ so that the equation of a straight line in this position will be

$$Y = 2X + 3$$

One common characteristic feature of the equations so far obtained, namely:

$$Y = X, \quad Y = 2X, \quad Y = -X, \quad Y = 2X + 3$$

is that every one of them is a linear equation with two unknowns. It was to be expected that there would be two unknowns, since points are characterized by two numbers. The fact to be stressed is that the equation of a straight line in any position is a *linear* one. Conversely, it can be shown that any linear equation with two unknowns, expressed in any form whatever, can be regarded as the equation of a certain definite straight line. Linear equation and straight line are two different expressions of the same concept.

This is a neat but not a very surprising result. We may draw straight lines in any sort of position; these will after all still be straight lines, they will belong to the same family, and it is quite natural that their equations should also form a definite family of equations.

Let us now have a look at a curved line. Everyone knows what a circle is. Let us, for example, consider a wheel with a lot of equal spokes; these spokes are the radii of the circle.

Let one of these radii be, for instance, 5 units long, and imagine the centre of the circle to be our starting point. Wherever we pick a point on the circumference of the circle and draw its co-ordinates and the corresponding radius passing through the point, we obtain a right-angled triangle. The hypotenuse

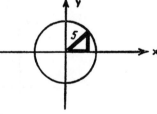

is the radius and the other two sides are the co-ordinates.

Let us remember the connexion we already know of between the sides of a right-angled triangle. This is good old Pythagoras' theorem. The square on the hypotenuse is equal to the sum of the squares on the other two sides. So if we square and add the co-ordinates of any point lying on the circle, we must get $5^2 = 25$:

$$X^2 + Y^2 = 25$$

This will be the equation of the circle.

We can see straight away that this is a quadratic equation; what is more, it is not the simplest kind of quadratic equation. Let us see what kind of a curve corresponds to the simplest kind, i.e. to the equation

$$Y = X^2$$

$$\text{If } X = -3 \text{ then } Y = (-3)^2 = +9$$
$$\text{If } X = -2 \text{ then } Y = (-2)^2 = 4$$
$$\text{If } X = -1 \text{ then } Y = (-1)^2 = 1$$
$$\text{If } X = 0 \text{ then } Y = 0^2 = 0$$
$$\text{If } X = 1 \text{ then } Y = 1^2 = 1$$
$$\text{If } X = 2 \text{ then } Y = 2^2 = 4$$
$$\text{If } X = 3 \text{ then } Y = 3^2 = 9$$

Let us take some intermediate values around 0.

$$\text{If } X = \frac{1}{2}, \text{ then } Y = \left(\frac{1}{2}\right)^2 = \frac{1}{4}$$

$$\text{If } X = -\frac{1}{2}, \text{ then } Y = \left(-\frac{1}{2}\right)^2 = \frac{1}{4}$$

Let us now draw the graph:

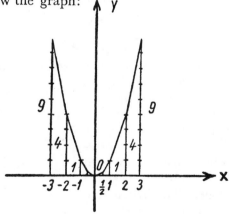

The curve obtained when this is quite smoothed out is called a parabola. Both sides of course continue indefinitely getting steeper and steeper, becoming more and more like vertical straight lines. This is certainly not even remotely like a circle.

We have come across another curve before whose equation is a quadratic, but we did not notice this fact. I am thinking of the hyperbola. Its equation was

$$Y = \frac{12}{X}$$

but if we bring X over to the left as a multiplier, we get

$$X \times Y = 12$$

In an equation with two unknowns the term $X \times Y$, in which the sum of the exponents of the two unknowns is 2, is usually regarded as a quadratic one. If this sounds unconvincing, we need only to turn our hyperbola around a little so as to bring it into the following position:

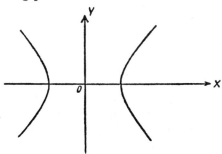

then its equation is going to be

$$X^2 - Y^2 = 24$$

and there can be no doubt about this being quadratic.

We might as well mention here that the equation of a compressed circle, i.e. of an ellipse,

is also a quadratic one. With it we have exhausted these types of curves (disregarding some 'degenerate' cases): if we could draw all the four types of curves discussed above in every possible position in our co-ordinate system, then we should obtain the family corresponding to all quadratic equations with two unknowns.   But it seems difficult to imagine a family whose members differ from one another so much.   Where is the family likeness in this family of curves, of which some are finite and closed, some wander off to Infinity, some are in one piece, others in two?

When this family is introduced it is immediately revealed where the family likeness lies: they all bear the name of 'Conic Sections'.

Here again we must leave the plane for three-dimensional space; what a pity it is that we cannot draw in three dimensions in the way we can draw on a flat piece of paper!   Let us at least imagine some paint with which the air can be painted.   Then let us imagine a horizontal disc and a slanting straight line leaning over the centre of the disc which just touches the disc at a certain point:

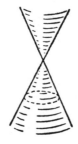

We next have to imagine that someone has dipped the straight line in the magic paint from top to bottom (it actually has no top or bottom, since the straight line is infinitely long).

Now let us take this imaginary straight line, holding it fast at the point which lies just above the centre of the disc with one hand and with the other hand exactly at the point where it touches the disc, and take this point round the circle.   Then the paint will paint a surface in the air below as well as above the fixed point.   Such a surface is called a cone.

If we cut the double cone so formed by means of planes in various positions, our curves will appear along the edges of the truncated pieces.

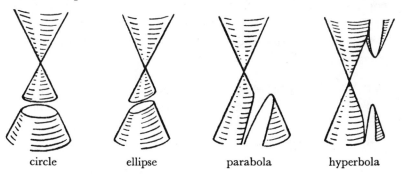

circle   ellipse   parabola   hyperbola

Only the fourth plane has succeeded in finding the upper cone as well.

Even had we been unable to find such a geometrical family likeness between the four curves, the very fact that all their equations are quadratic brings to light a number of their common characteristics. We merely need to ask what Algebra has to say about such equations, and establish what can subsequently be deduced from this; anything that we may deduce in this way will be a common property of our four curves. Let us for example have a look at their points of intersection with a given straight line. A point of intersection is a point which is on the curve as well as on the straight line, so its co-ordinates satisfy both the equations in question. The equation of a straight line is linear; Algebra teaches us that a linear and a quadratic equation, each with two unknowns, either have no (real) solution, or one common solution or two. Thus it is true for any of our conic sections that a straight line can be in one of three relationships with respect to it: either it does not get anywhere near it, or it touches it in one point, or it cuts it in two points, for example:

No line can cut even the two-piece hyperbola in more than two points.

Such are the services that Algebra can render Geometry.

Postscript about waves and shadows

During our discussions we have come across two geometrical ideas. It would be a pity to pass them by.

One is to do with different ways in which the direction of a straight line can be given: we can compare the rise in height with the horizontal distance travelled, as in the right-angled triangle drawn below, where the two sides adjacent to the right angle are compared:

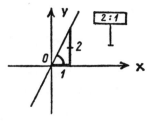

There is, of course, another way of precisely determining a direction; we can state what angle it makes with a certain definite direction. It is usual to take the positive direction of the X-axis for such a definite direction. This angle may be called the direction-angle of the straight line. It is an acute angle if the straight line rises towards the upright, it is obtuse if it falls away beyond it:

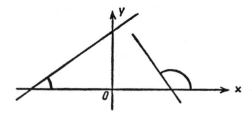

Now the ratio between the two sides of the right-angled triangle adjacent to the right angle completely determines the direction, and so with it the angle too. We could therefore choose this as a measure of our angle. We could for example describe the size of the acute angle by saying that the corresponding slope is

2 : 3; i.e. if we drop a perpendicular from any point on one of the arms of the angle on to the other arm, we shall get a right-angled triangle in which the ratio of the side opposite the angle to that adjacent to the angle is 2 : 3, or with a different notation $\frac{2}{3}$.

If given the ratio $\frac{2}{3}$ the angle can immediately be drawn. We must go 3 units to the right and 2 units upwards:

and, if we join the point we have reached to our starting point, we shall obtain the required angle:

If the angle is obtuse, then the slope falls away, and we have already seen that the corresponding ratio is going to be negative. If for example the ratio is $-\frac{2}{3}$, then we know that the slope rises backwards, i.e. if we go 3 units to the left and 2 units upwards, then join the point so obtained to our starting point, the join will make an obtuse angle with the positive direction of the X-axis (since the direction-angle is always the angle made with the positive direction of the X-axis):

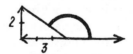

An obtuse angle like this cannot form a part of a right-angled triangle, but still we have constructed a right-angled triangle right next to it, whose sides adjacent to the right angle are in the ratio of $\frac{2}{3}$; this is the ratio characterizing our obtuse angle, apart from the sign.

It can be shown that the ratio of any pair of sides of the right-angled triangle can characterize angles as well. These ratios are called circular functions, since their values depend on the amount of circular movement of one arm of the angle away from the other. The name given to the circular function we have just examined is the tangent; the ratio of the side opposite the angle to the hypotenuse is the sine of the angle, the ratio of the side adjacent to the angle to the hypotenuse is the cosine of the angle.   For example in the triangle below:

the sine of the shaded angle is $\frac{3}{5}$, its cosine is $\frac{4}{5}$.   The definitions of all circular functions can be extended to angles larger than acute angles.   The values of the circular functions corresponding to all sorts of different angles have been tabulated; if we know the lengths of the sides of a right-angled triangle (and other triangles can always be split into two right-angled triangles):

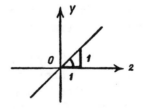

we have only to look in the tables and we immediately know all its angles as well.   It is true that we can draw the triangle if we know the lengths of its sides, and then we can measure its angles, but the standard of accuracy of such measurements is far below what can be achieved through the calculations of those who prepare the tables!   For it must not be imagined that the compilers of such tables obtain the values of the circular functions by measurement!   One method for their calculation is based on the fact that we know some of the values precisely.   For example, our first straight line bisected the right angle exactly

and so the tangent of its direction-angle was $1 \div 1$, i.e. $\frac{1}{1} = 1$.

The right angle is the result of a quarter turn; so we know that the tangent of the angle which is the result of an eighth of a turn is 1.   If we know the values of the circular functions for some angles, we might enquire how we can find the values corresponding to the sum of two such angles, or to double or half an angle.   Trigonometry is concerned with the search for such relationships.   The tables, on the other hand, are prepared in a different way; we shall come back to this later on.

The circular functions have great importance far beyond the bounds of Trigonometry.   If, for example, we draw the chart corresponding to the sine function for all values of the angle starting from *O* right up to one complete turn, we get a wavy line like this:

This can be continued even farther.   An angle really measures the turning away of one straight line from another fixed straight line.   Imagine, for instance, that we slowly open out a Japanese fan:

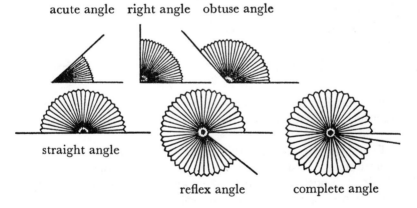

acute angle   right angle   obtuse angle

straight angle

reflex angle   complete angle

in this way we can make all possible angles, and it is easily seen that the arc of a circle drawn about the vertex measures the amount of turning.   Of course the length of the arc also depends on the size of the fan, but we can measure our angles by the length of the arc of unit radius.   (This is much more sensible than the usual degrees used in schools.)   Now we can imagine (not with the fan, this would get torn) that the turning straight line goes on turning even after one complete revolution

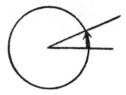

so that the part thickly drawn in the figure has been covered twice.   It is quite obvious that our straight line is in just the same direction as if it had turned only through this small arc.

So the values of the circular functions get repeated for angles greater than one complete turn, and the curve goes on rising and falling in the same way:

This is just like the recurrence of the periods in the expansion of fractions.   For this reason the sine function is called a periodic function.

Every physicist knows this curve very well: it is the curve representing vibrations and plays a decisive role in modern Physics.   Those who have taken an interest in radio might have seen such modified wave-graphs as this one:

In this graph the denser waves are the so-called electro-magnetic waves. Their picture by itself would be like this:

but sound modifies these in places by means of big waves like these:

Here we can still see the two sets of waves out of which the modified wave is compounded. But in fact sound waves are never so simple, as there is no such thing as absolutely pure sound; there are always several sounds vibrating at the same time, and these do not differ from one another to the extent that electromagnetic waves differ from sound waves, and so they do not play such easily distinguishable parts. The results of their superimposition is merely a degeneration of the waves, for example they might become like this:

It is often necessary to discover, given such a degenerate set of waves, what the waves were, out of which our curve has been compounded. The question can be put in the following way: if we have a continuous curve, however degenerate, but periodic, is it possible to find waves whose simultaneous effect would generate just this curve?

The answer to this question is that it is possible to find such waves (although not with perfect accuracy), which, if super-imposed upon one another, approximate to our curve to any desired extent. This can be done even if our curve consists of a lot of elbows, for example if it consists of a lot of segments like this:

This is of course proved in function-language, which deals not with waves, but with the functions corresponding to these.

In this field pioneer work was done by Lipót Fejér and this made his name as a young man.

   *   *   *

The other geometrical idea we have come across is in connexion with slicing up the cone. Let us just cut the lower cone twice; once in a horizontal plane, once in a slanting one:

Let us draw separately the vertex of this cone, the circle and the ellipse.

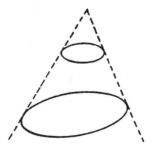

Let us imagine that the vertex of the cone is a little electric light bulb which emits rays of light in all directions. The circle is a paper disc, interrupting the rays of light, which therefore does not allow the rays falling on it to pass through it; the rays just able to get past the edge generate the surface of our cone, and so the circle produces an elliptical shadow on the plane put under it in a slanting position:

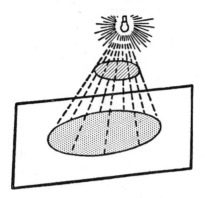

An ellipse can therefore be regarded as the shadow of a circle. It can be generated by the projection of a circle from a point on to a slanting plane.

In the same way we can produce a parabolic or a hyperbolic shadow by turning the plane about (if we wish to obtain the other branch of the hyperbola, we need to place an identical circular disc in the way of the rays emitted upwards). We can see that the extent to which a shadow can distort is considerable.

So-called 'Projective Geometry' has for its object the search for properties which are not lost even through the distortion caused by projection. It has been possible to find such 'projective' properties which have proved 'invariant' even under such projections. This enables us to examine conic sections in a uniform and simple way from quite a novel angle. It is enough to deal with the well-known circle; all its 'projective' properties will be transmitted unscathed to all conic sections that can be generated out of it by projections. The shadow can stretch and stretch, even to Infinity. Yet it cannot altogether free itself from its master.

15. 'Write it down' elements

I ONCE went to see the stage version of an amusing Russian short story called 'General, write it down!' The main idea behind the story is that someone misunderstands an interjection 'General, write it down!' in the middle of something being dictated, and so the name 'General Writeitdown' gets put on the list of officers. Since the omnipotent Czar signs this list, nobody dares to come forward with the information that there is in fact no General whose name is Writeitdown. General Writeitdown is therefore not a human being at all, he is only a misprint, but nevertheless all sorts of extraordinary things happen to him and about him: he gets imprisoned, gets married, foments insurrection and generally has a decisive influence on the lives of other people.

Even in Mathematics we find such non-existent 'writeitdown' elements, which nevertheless play an important part. Mathematicians call them ideal elements. Such is for example the so-called 'point at infinity', in which 'parallel lines meet'. This serves the purpose of making our discussions more unified. For example it can be proved that points and straight lines are in a relationship of 'duality' to each other: some theorems concerning points and lines remain true if we interchange the words 'point' and 'straight line'. For example, 3 points, not lying on the same straight line, determine a triangle. This is certainly true:

The dual statement would be: 3 straight lines, not passing through the same point, also determine a triangle:

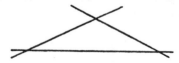

This duality is very convenient; it is enough to prove one of the statements, and we have automatically proved its dual. We state one thing, and it immediately becomes two things.

But in this very simple example something is a little wrong with the dual theorem. We ought to have added: 'provided the straight lines are not parallel'. In this case it is convenient to say that we have already excluded the case of three parallels in the wording of the theorem, since these parallels would meet in one and the same point at infinity.

But this ideal point at an infinite distance away is capable of greater things than merely saving a few sentences beginning with 'provided'. If we associate a single common point at infinity with straight lines having the same direction, i.e. with lines parallel to each other, and a different point at infinity to straight lines with different directions, then we have created as many ideal points as there are directions. We can even state exactly which ideal point we are talking about; we have only to give the direction which points towards it. By a little modification of our co-ordinate system, we can even write down the equation of the line containing all the ideal points. This equation turns out to be the kind of equation usually associated with straight lines, and so we can say that all the points at infinity lie on a straight line at infinity.

So far this may appear a very empty sort of game: we have written down the equation of a non-existent straight line. Perhaps it would be better not to try even to imagine it. A straight line is infinite both ways, and yet we have associated only one point at infinity with it (in this way the duality principle is satisfied, two ideal points would spoil it); it is as though its two ends met at infinity, where it would turn into some sort of a circle. Our straight lines, although stretching away in two opposite senses, are nevertheless hanging on the various points of the line at infinity, like fruit on a fruit-tree, turned into circles by magic, parallel lines hanging at the same point:

We should not have drawn the line at infinity so straight, really, although goodness knows how we could have drawn it, since one of its points is in the East and the West at the same time;

North and South meet at one of its other points and so on for all other directions.   Let us rather forget the whole thing, it does not belong to a world of imaginable things.   'General Write-itdown' is only a misprint.

But nevertheless this line at infinity can give us an enormous amount of information.   We have already obtained its equation, and so it is perhaps not too daring an enterprise to try to determine its points of intersection with, for example, a parabola, since all we have to do is to find the common solutions to the two equations.   It happens then that this line at infinity, which we at first thought was trouble personified, is just what we wanted to throw light on the family of conic sections.

Everyone who has had anything to do with this subject is bound sooner or later to ask the following question: given a quadratic equation with two unknowns, how is it possible to decide what kind of conic section corresponds to it?   The line at infinity gives a definite answer to this question: if the given equation has no common solution with the line at infinity, then it is an ellipse; if it has only one common solution, it is a parabola; and if it has two solutions, then our equation must represent a hyperbola.   And there are no other possibilities.   (The circle is the most regular sub-case of the ellipse.)

Now we can allow our imagination free rein—and the results obtained correspond entirely to our imagination.   The ellipse lies entirely within a finite region, so of course it has no common point with the line at infinity.   The two sides of the parabola get steeper and steeper, they become more and more like two parallel straight lines, so it is quite natural that they meet in one and the same point at infinity.   The two branches of the hyperbola stretch away into the distance along two asymptotes with different directions; it is likewise quite natural that these branches reach Infinity at two distinct points.

Perhaps the reader will now agree that it would have been a pity not to let these non-existent points speak.

Now I can pluck up courage to come back to the last of our problems which had been temporarily shelved, namely to quadratic equations like this one:

$$X^2 = -9$$

I should denote a number whose square was -9 by the sign

$\sqrt{-9}$, if such a number existed. The trouble is that we have not yet come across numbers whose squares are negative. Whether we square -3 or $+3$, the result is always $+9$. We have even no idea of what $\sqrt{-1}$ might be. 'I don't know' is the only truthful answer. But let us suppose that, since I was thinking very hard, I rather drawled over the 'I', and someone who was taking notes feverishly of everything I was saying thought that i was perhaps the answer, i.e.

$$\sqrt{-1} = i$$

and so he excitedly interrupts me by saying: 'Then I also know what $\sqrt{-9}$ is, it must be $3i$, or it could also be $-3i$!' Well, in this he is of course quite right: if $\sqrt{-1} = i$, then i is the number whose square is -1,

$$i^2 = -1$$

and so

$$(+3i)^2 = 3i \times 3i = 9i^2 = 9 \times (-1) = -9$$

or

$$(-3i)^2 = (-3i) \times (-3i) = 9i^2 = 9 \times (-1) = -9$$

The only trouble is that this i does not exist, the whole thing is a misunderstanding, a misprint. In actual fact we do not know what $\sqrt{-1}$ is.

But now that the misprint has got into the book, let us play about with it a little, just as we did when we worked out what $\sqrt{-9}$ was. Perhaps this non-existent element can do a thing or two as well.

We shall see that it can do quite amazing things. The whole of Function Theory, which is the most respectable branch of Mathematics, is based on it. If this i is to be left out, it has to be specially stated that Real Function Theory is meant. There is no branch of Mathematics which does not turn to this i for help, especially when something of a deep significance needs to be expressed; even Geometry is no exception to this. The attempts at systematic unification of apparently quite independent theorems are crowned with success thanks to i.

I can give the reader only a taste of such unifications in this Mathematics without formulae, since ideal elements exist entirely in virtue of their form.

For example, if we allow the use of i, connexions previously undreamed of suddenly appear between certain functions.

Who would imagine that there was any connexion between the circular functions and the power function?

Nevertheless it can be proved, if we measure angles by the length of the arc of a circle of unit radius drawn about the vertex,

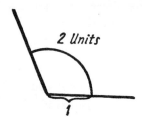

that the cosine of an angle of two units (written for short cos 2) can be written as follows:

$$\cos 2 = \frac{e^{2i} + e^{-2i}}{2}$$

where e is the base of the natural logarithms. A similar formula holds for angles of any size:

$$\cos 3 = \frac{e^{3i} + e^{-3i}}{2}$$

$$\cos 4 = \frac{e^{4i} + e^{-4i}}{2} \quad \text{and so on.}$$

How is it possible that the cosine of an angle, which is after all the ratio of two numbers and so an honest-to-goodness real number, can be equal to the non-existent number on the right?

This is possible only if the number on the right is also a real number. While the operations designated on the right are being performed, i suddenly appears from some imaginary world, throws light on the relationships, and then disappears again. This sort of thing can occur in the games we play of finding out a number that somebody has thought of. For example: 'Think of a number, multiply it by 3, add 4 to it, then multiply what you have by 2, and subtract from it 6 times the number you thought of.' Here we can wait till our friend has finished the problem, we need ask no further questions, yet we can say: 'The result was 8!' In fact, we can write out the steps as follows: Let the number be X, this multiplied by 3 is $3X$, adding 4 it will become $3X + 4$. This must be multiplied by 2, so we have $2 \times (3X + 4)$; and finally we must subtract 6 times the number we thought of, and so

$$2 \times (3X + 4) - 6X$$

will be the result. Multiplying both terms in $3X + 4$ by 2, we shall have

$6X + 8 - 6X$ or, written in a different order, $8 + 6X - 6X$

but if we add $6X$ to 8 and then subtract $6X$ afterwards, surely we shall be left with 8. The number we thought of came into our calculations, but it disappeared again.

Out of the connexion between the circular functions and powers it is even possible to derive relationships in which there is no apparent trace of i. For example let us calculate the square of cos 2 from the formula

$$\cos 2 = \frac{e^{2i} + e^{-2i}}{2}$$

and to avoid the bother of fractions let us take the divisor 2 from the right over to the left as a multiplier:

$$2 \times \cos 2 = e^{2i} + e^{-2i}$$

Now let us square this equation. The square of the left-hand side is

$$2 \times \cos 2 \times 2 \times \cos 2 = 2 \times 2 \times (\cos 2)^2$$

(there is a good reason for pretending to forget that $2 \times 2 = 4$).

The right-hand side is the sum of two terms and to square it we must first square the first term, remembering that when we square a power we multiply the exponents:

$$(e^{2i})^2 = e^{4i}$$

To this we must add double the product of the two terms, not forgetting that we can multiply the powers of e by adding the exponents, and that the value of the 0th power is 1:

$$2 \times e^{2i} \times e^{-2i} = 2 \times e^{2i+(-2i)} = 2 \times e^0 = 2 \times 1 = 2$$

Finally we must add the square of the second term:

$$(e^{-2i})^2 = e^{-4i}$$

so that the square of the right-hand side can be written as follows:

$$e^{4i} + 2 + e^{-4i}$$

or, in a different order,

$$e^{4i} + e^{-4i} + 2$$

and so $2 \times 2 \times (\cos 2)^2 = e^{4i} + e^{-4i} + 2$

Now let us take one of the multipliers over to the right as a divisor, one of the 2's. All the terms on the right-hand side must then be divided by 2. We can divide 2 by 2, that will be 1, and we can just indicate formally the division of the other two terms:

$$2 \times (\cos 2)^2 = \frac{e^{4i} + e^{-4i}}{2} + 1$$

But here we have come across an old friend

$$\frac{e^{4i} + e^{-4i}}{2}$$

which is just the one we said was equal to cos 4. So let us replace it by cos 4:

$$2 \times (\cos 2)^2 = \cos 4 + 1$$

or, in a different order (since someone might think that we are talking about the cosine of 4 + 1, i.e. the cosine of 5):

$$2 \times (\cos 2)^2 = 1 + \cos 4$$

Finally let us take the second 2 over to the right as a divisor:

$$(\cos 2)^2 = \frac{1 + \cos 4}{2}$$

This is one of the well-known trigonometrical relationships; and there is no trace of i in it. It may be quite comforting to know that we have not made a mistake in our calculations; but this is nothing new. But if we remember that the sum of two terms can not only be squared, but can also be raised to any power with the aid of the binomial theorem, then at one stroke we can prove a whole lot of trigonometrical theorems.

I beg the reader's pardon for the lengthy calculation in which he needed to remember so many different rules all at once. But I think this is unavoidable; in order to understand, the reader needs to experience for himself at least once how this i disappears from the calculations, having infused them with a new lease of life.

But we shall see that this is not the most important part that it plays.

It is quite natural that it should provide solutions to the insoluble cases of the quadratic equation, since it was really introduced to deal with these situations, enabling us to extract square roots of negative numbers. It is true that we obtain only an 'imaginary' result, but, after the foregoing, perhaps the reader will be convinced that such imaginary results are not to be lightly cast aside. For example the solution of

$$(X - 2)^2 = - 9$$

is
$$X - 2 = \sqrt{-9}$$

and $\sqrt{-9} = 3i$, or $\sqrt{-9} = -3i$. Let us now take the subtracted 2 from the left over to the right as an added 2, and we shall obtain the two 'roots' (the solutions of equations are called roots, since often we obtain them after the extraction of roots):

$$X = 2 + 3i$$

or
$$X = 2 - 3i$$

These are numbers consisting of a real and an imaginary part. This strange kind of joining up of the real and of the imaginary worlds is known as 'complex number'. These numbers might appear rather impossible at first sight, although their sum is real; when we add them up the $3i$ and the $-3i$ cancel each other out. Moreover, it can quite easily be seen that their product is also real.

Among the complex numbers are also to be found the real and the purely imaginary numbers. For example $5 + 0i = 5$ is real and $0 + 2i$ purely imaginary.

If we want to take the fourth root, or the sixth or the eighth root, of a negative number, we get stuck in the same way as we did in the case of square roots. If we raise a positive or a negative number to an even power, we are bound to get a positive result. For example the fourth root of -16 is neither positive nor negative, since

$$(+\,2)^4 = 2 \times 2 \times 2 \times 2 = 16$$

and $\qquad (-\,2)^4 = \underbrace{(-\,2) \times (-\,2)} \times \underbrace{(-\,2) \times (-\,2)}$

$$= (+\,4) \times (+\,4)$$

which is also $+16$. We might wonder if we now need to introduce new ideal elements. It turns out, rather oddly, that this is quite unnecessary. We can carry out all these operations with the aid of what we already have. Moreover, it can be proved that any equation of any degree can be solved within the field of complex numbers. This is called the fundamental theorem of Algebra. It does not contradict Abel's result that we are bound to get stuck with the solutions of equations of the fifth degree; the fundamental theorem is proved in the sense of 'pure existence'; it gives no method for finding any number satisfying the equation by means of the fundamental operations and extraction of roots.

The extraction of square roots always gives two values, a positive and a negative one. This is why a quadratic equation always has two roots in the field of complex numbers. Or rather, not quite always, for the equation

$$(X - 3)^2 = 0$$

has only one solution, since the number whose square is zero can only be zero itself, so that

$$X - 3 = 0$$

or $\qquad\qquad\qquad\qquad X = 3$

is the only solution. The expanded form of this equation is

$$X^2 - 6X + 9 = 0$$

We can find equations which approximate to this one more and more, i.e. in which the numbers differ from the above 6 and 9 by less and less. Each of these will have two roots, but these roots will get nearer and nearer to each other as the equations become closer and closer to our one. This is why it is sometimes said that the moment these equations become exactly identical with the equation

$$X^2 - 6X + 9 = 0$$

the two roots 'coincide'.

How many roots would an equation of the fourth degree have? We can solve the equation

$$X^4 = 1$$

without the aid of i. We can raise either $+1$ or -1 to the fourth power; we shall get $+1$, so it would appear that it has two roots, $+1$ and -1. But now i interrupts: 'Oh, no! this is quite irregular, the equation is of the fourth degree, it ought to have four roots; here am i.' And in fact i is a root, and even $-i$ is a root, since

$$i^4 = \underbrace{i \times i} \times \underbrace{i \times i} = i^2 \times i^2 = (-1) \times (-1) = +1$$

$$(-i)^4 = \underbrace{(-i) \times (-i)} \times \underbrace{(-i) \times (-i)}$$
$$= i^2 \times i^2 = (-1) \times (-1) = +1$$

In this way i tidies up the roots of all the equations. It can be proved that within the field of complex numbers equations have as many roots as their degree, apart from some roots which may coincide.

Such is the service that i renders Algebra.

Its greatest service, however, is reserved for Function Theory.

But in order to give the reader even a small taste of this, I must explain the graphical representation of complex numbers.

Let us consider i as a new kind of unit; the multiples of i can be represented on a new line. The zero point of this line may coincide with the zero point of the real line, since $0 \times i = i$. In this way the two lines can be considered as being similar to a co-ordinate system

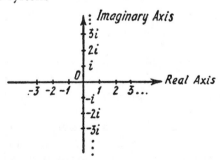

This suggests that the complex numbers can be represented by means of the points of the plane, as they consist of real and

imaginary parts; the X co-ordinate will be the real part, the Y co-ordinate the imaginary part. Below will be found the pictures of a few complex numbers:

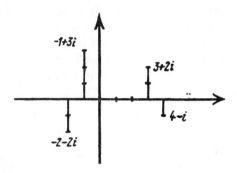

Complex numbers are therefore distributed not along a line but all over a plane.

By the absolute value of a complex number is meant its distance from the zero point. This distance can be small or large, but there are a whole lot of complex numbers, all at the same distance from the zero point; these are situated on a circle whose centre is at 0:

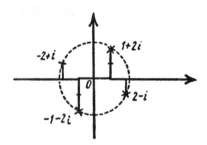

There is absolutely no reason why we should call one of these smaller than the others. There can therefore be no question of any kind of 'less than' or 'greater than' concept with complex numbers.

Nevertheless we can easily assure ourselves that all the old rules of manipulation have remained intact, as long as we treat the complex numbers as ordinary numbers and i as some unknown about which all we know is that, every time i^2 occurs, we can replace it by -1.

Now let us go back to one of our previous results. We obtained from the chocolate example that

$$1\frac{1}{9} = 1 + \frac{1}{10} + \frac{1}{100} + \frac{1}{1000} + \cdots$$

Here on the right-hand side every number is $\frac{1}{10}$ times as big as the previous one; $\frac{1}{10}$ is the 'common ratio' of the geometrical series. Let us try to transform $1\frac{1}{9}$ in such a way that there is a $\frac{1}{10}$ in it.

$$1 = \frac{9}{9}, \quad 1\frac{1}{9} = \frac{10}{9}$$

We have done quite a lot of simplifying. We know that we can divide numerator and denominator by the same number; let us divide here by 10, not bothering about the fact that the division can only be indicated in the denominator:

$$1\frac{1}{9} = \frac{1}{\frac{9}{10}}$$

This is useful because we can express $\frac{9}{10}$ quite easily in terms of $\frac{1}{10}$: one whole consists of 10 tenths, and if we take away one tenth, we are left with exactly $\frac{9}{10}$, so that

$$\frac{9}{10} = 1 - \frac{1}{10}$$

Finally we have

$$1\frac{1}{9} = \frac{1}{1 - \frac{1}{10}}$$

If we replace $1\frac{1}{9}$ by this, we have

$$\frac{1}{1 - \frac{1}{10}} = 1 + \frac{1}{10} + \frac{1}{100} + \frac{1}{1000} + \cdots$$

In this form our result can be generalized. If instead of $\frac{1}{10}$ the common ratio of the geometrical series is for example $\frac{2}{3}$, then every term is $\frac{2}{3}$ times the previous one, so that the terms of the series will be 1, 1 × $\frac{2}{3}$ = $\frac{2}{3}$, $\frac{2}{3}$ × $\frac{2}{3}$ = $\frac{4}{9}$, $\frac{4}{9}$ × $\frac{2}{3}$ = $\frac{8}{27}$, and so on, one after the other, and we can prove that

$$\frac{1}{1 - \dfrac{2}{3}} = 1 + \frac{2}{3} + \frac{4}{9} + \frac{8}{27} + \cdots$$

But we must be careful, for we have seen that not every geometrical series can be summed. For example when the common ratio was 1 or −1 or any number whose absolute value is even greater, the series could not be summed. It can be proved that if the common ratio is nearer to zero than 1, the series is convergent, and its sum can be expressed in the same way as in the case of the values $\frac{1}{10}$ and $\frac{2}{3}$. So all the common ratios for which the series can be summed lie between −1 and +1 on our line

If we think of one of these numbers, without saying which one, we can then call it X. Even without knowing what the number is, we can say that the terms of the geometrical series constructed from it will be

1, 1 × $X = X$, $X \times X = X^2$, $X^2 \times X = X \times X \times X = X^3$, $X^3 \times X = X \times X \times X \times X = X^4$, …

and that it is likewise true for this series that

$$\frac{1}{1 - X} = 1 + X + X^2 + X^3 + X^4 + \cdots$$

This will always be true, whatever X is, provided I am careful to choose it from the interval of numbers ranging from −1 to +1.

The value of $\dfrac{1}{1 - X}$ naturally depends on what number X really is, so it is a function of X. It is usual to express the above relationship by saying that we have expanded the function in a power series, or else in an infinite series consisting of ever-increasing powers of X. It is the partial sums of this series that give a better and better approximation to the value of $\dfrac{1}{1 - X}$.

As a first, rather rough approximation we can actually replace $\frac{1}{1-X}$ by 1; $1 + X$ is a better approximation, $1 + X + X^2$ an even better approximation and so on. The question arises whether it is possible in general to expand a function in a power series (of course we could not expect a series like the previous one, but one in which the powers of X are multiplied by certain numbers)? This is a question of fundamental importance in the Theory of Functions. The function $\frac{1}{1-X}$ is still quite simple, its values can easily be calculated. But it has also been possible to expand the power function as a function of the exponent in a power series; what is more, this has turned out to be simplest when the base is $e = 2\cdot71 \ldots$ The series is as follows, where X can be any number:

$$e^X = 1 + X + \frac{1}{2!}X^2 + \frac{1}{3!}X^3 + \frac{1}{4!}X^4 + \ldots$$

where—perhaps the reader will not have forgotten—

$$2! = 1 \times 2, \quad 3! = 1 \times 2 \times 3, \quad 4! = 1 \times 2 \times 3 \times 4$$

and so on.

This is a great help in the calculation of the values of e^X, if we write definite numbers in place of X. It would not be very amusing to raise e, this irrational infinite decimal, to different powers. If X is small, $1 + X$ is going to be a quite good approximation for it, and to work this out, i.e. to add a number to 1, is really child's play. If we require greater accuracy, we can take a longer partial sum; in this case we have to calculate some of the powers of the given number. For example if $X = \frac{3}{10}$, it is still much easier to raise this number to the second, third, fourth powers, than to extract the 10th root of the irrational number $(2\cdot71\ldots)^3$, which is after all what $(2\cdot71\ldots)^{\frac{3}{10}}$ really means.

It is very fortunate that this expansion is true for all values of X.

The circular functions as well as the logarithm function can be expanded in power series, and nowadays their tables are prepared on this basis.

These series, however, are not all convergent for all values of

X, and great care must be exercised lest we try to replace something by a supposed approximation when there can be no question of approximations at all. So the problem arises, given a function, how it can be ascertained for which values of X it may be expanded in a power series.

Let us have another look at our geometrical series. We have stated that the expansion

$$\frac{1}{1-X} = 1 + X + X^2 + X^3 + X^4 + \dots$$

is valid in the interval between -1 and $+1$

Is it possible to see from $\frac{1}{1-X}$ that 1 is this limit of validity (it will be at the same distance on the other side of 0)?

It hits you in the eye! What would happen if X stood for 1?

$$\frac{1}{1-1} = \frac{1}{0}$$

it is even painful to write it! Here is the perpetual prohibition, division by 0! Even if we do not look at the series, the function itself calls 'Halt!' at the point 1.

Does the function always give away in such a definite way the limits to which we can go?

If we are dealing with real numbers only, this is not so. This fact resulted in a lot of trouble with a number of functions. It was time that i intervened, and by so doing it cleared the whole question up once and for all.

Let us take an example.

Those who are a little handy with formulae will see straight away from our geometrical series that the function $\frac{1}{1+X^2}$ can be expanded in the following power series:

$$\frac{1}{1+X^2} = 1 - X^2 + X^4 - X^6 + \dots$$

and this series is also convergent if and only if X lies between -1 and $+1$.

Does this function give away the secret of the limits beyond which its expansion will not be valid?

Let us write 1 for X:

$$\frac{1}{1 + 1^2} = \frac{1}{1 + 1} = \frac{1}{2}; \text{ no trouble.}$$

Perhaps the trouble is at the other limit.   Let us write -1 for X,

$$\frac{1}{1 + (-1)^2} = \frac{1}{1 + 1} = \frac{1}{2}; \text{ no trouble here either.}$$

Now we are in a bit of a mess.

This is where i comes to the rescue: 'Why don't you write me in for X?' Let us try:

$$\frac{1}{1 + i^2} = \frac{1}{1 + (-1)} = \frac{1}{0}$$

Halt! This is a division by zero. If we think of the complex plane, we see straight away that i lies at a unit distance from the 0 point, and the fact that there is trouble with the function at such a point indicates that it is forbidden to go beyond the unit distance from 0

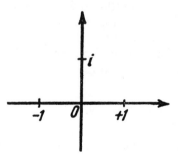

So there is some point in examining the values of a function at complex points, not merely at the real points. And it is quite generally true that if there is some trouble with the function at any one point, then the function cannot be expanded in a power series at any point which is farther from 0 than the troublesome point. So we have to find the nearest troublesome point to the zero point in the complex plane. This is as far as the circle will reach inside which the function can still be expanded in a power series.

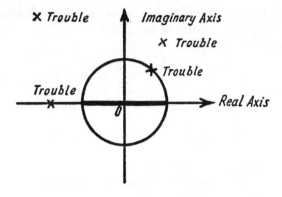

In this way we obtain a circle inside which the series will be convergent, and perhaps also at some points on its circumference, but outside which it certainly will not be. Such a circle will always cut an interval about the zero point out of the real axis, shown by a thick line on the figure.

So *i* has come along again, put everything in order, and, if we wish it, can discreetly disappear. We may restrict ourselves to this real interval of numbers which was exactly determined by it. But the bewitched mathematician will not let it go now. If it can do all these things, it cannot really be called non-existent. It is worth while to explore the Theory of Complex Functions, this 'World created out of nothing', which is much more orderly than the real world.

16. *Some workshop secrets*

AFTER recovering from the immediate effects of standing face to face with a masterpiece, it is natural to begin to wonder about more mundane things; one might wish to know how the master-piece came into existence, of what its essentially human element might consist, the detailed problems of the worker, the sweat of his brow. It would in short be interesting to have a look inside the workshop.

Let us come back from our imaginary world and see if we can discover some of the mathematician's workshop secrets. I mean the detailed, day-to-day work, which I really wanted to spare the reader, but which I cannot altogether hide from him. The writer who started me off on this book was actually curious to know about the differential coefficient, and the differential coefficient certainly belongs to the mathematician's technical store. Even though it is not as brilliantly alluring a subject as that we have already touched upon, its importance is extra-ordinarily great; there is never a masterpiece without a lot of fiddling detail.

We have emphasized from the beginning that the concept of function is the backbone of the whole of Mathematics, and it is the associated curve which gives us a picture of the function. But by the nature of things this picture is bound to be imperfect. We have constructed the curve out of straight segments and tried to smooth them out by using more and more of them, but the pencilled segments have already coalesced after the first few steps in the smoothing process; a polygon with 16 sides is hardly distinguishable from a circle when we draw it. Nobody will believe that we can possibly derive serious relationships con-cerning our functions from such rough pictures. We need more of a precision instrument, which will register small varia-tions, however fine; an instrument which can follow the be-haviour of a function to any desired degree of accuracy. The differential co-efficient is just such a precision instrument.

Let us begin with the picture.

When we tried to obtain a picture of the parabola, we said

177

that both its parts become steeper and steeper.   But how is it possible to speak about the direction of a smooth curve?   We know what is meant by the direction of a straight line, since we can check its rise at any of its points; we can be sure about it: it will never deviate from a direction once assumed.   But the curve is a curve just because it changes its direction.   If we get hold of it at one of its points, we may well ask: 'What is your direction just here?'

But the curve is smooth and slips out of our hands without giving any definite answer.   Somehow we still feel that the curve has some sort of definite direction even at this point; it was not meaningless talk to talk about the steepness of the parabola.

Let us run our film back to the point where our curve was not yet so smooth.   Let us choose one of its definite points on a picture of the curve in this state:

At the point indicated there was still an 'elbow'.   Here the curve had not yet any definite direction, because before the point its direction was like this:

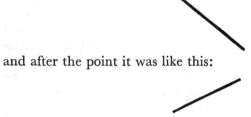

and after the point it was like this:

and so at the point itself there was a change of direction in our curve. Now let us turn the film forward to where we have taken several intermediate points:

Here the elbow is not quite so acute:

and the two directions that meet at our point hardly differ from one another.

We cannot go much farther by drawing, i.e. we cannot follow what will happen to the curve if we put in more intermediate points, but everyone can imagine that the elbow will get straighter and straighter, and the direction before the point will differ less and less from the direction after the point. By the direction of the curve at our point we ought to understand the common direction to which each side of the elbow approximates more and more as the elbow itself gets straighter and straighter.

When we are satisfied that these two directions approximate to the same common direction, then it will be sufficient to deal with one side only.

Let us take, for example, the segments *after* our point. We can see their directions better if we make them longer:

In this way we get different secants of the curve one after the other. As we put in more points, the nearest point gets nearer and nearer to our point, and smaller and smaller portions of the secant will fall inside the curve. We can observe what happens quite clearly if we replace the secant by a ruler and keep on turning it outwards while making sure that one of its points is permanently fixed over our point:

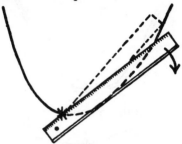

There will be a moment when the neighbouring point coincides exactly with our point, and the ruler will come away from the curve altogether.

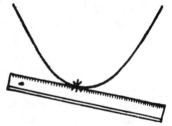

The secant has now become a tangent:

We feel that just at this moment we have caught the direction to which the upper part of the elbow approximates. If we approached the curve from the outside with a ruler placed along this direction, then it would reach the curve just at our point and

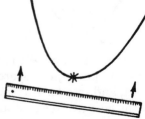

for one moment it would cling to the curve; if the two cling together, they have the same direction. We are in the lucky position of not having to examine this direction in the extremely small place where they cling together; the straight line preserves the memory of this moment for ever, its direction remains the same all the time.

Now we know what we should mean by the direction of a curve at one of its points: it is the direction of the tangent drawn to the curve at the point in question. This can be completely characterized by the ratio by means of which we express the rise of the slope. This will be the differential coefficient.

We have already come across the notion of tangent once before. We met it when we obtained the result in a purely algebraic way that a conic section may have 0, 1 or 2 points in common with a straight line, and we said that if they have only one point in common the straight line touches the conic section. This is true enough of conic sections, but in general the decisive property of a tangent is not that the straight line and the curve in question have only one common point. For example we may have a curve with an elbow left in it:

the straight line passing through it cannot really be considered a tangent, even though it has only one point in common with the curve. It is quite obvious that this straight line cannot indicate the direction of the curve. The curve has no ascertainable direction at this point, and our straight line does not even indicate the left- or the right-hand direction of the curve there.

On the other hand this straight line

has two points in common with the curve, yet we still have to regard it as a tangent *at the first point*, where it clings to the curve so nicely.

We might think that the decisive property would be that tangents touch and secants intersect, but even this is wrong. Take for example the straight line below

which manages to intersect the curve at the very moment when it clings to it.   In spite of intersecting it, it clings beautifully to the lower as well as to the upper part and there is no reason why we should not regard it as a tangent.

The only decisive condition is whether we reach the straight line in question at the moment when the ever-approaching secants, passing through neighbouring points, leave the curve. This is so in the last two cases, and the reader is advised to verify this by using a ruler turning about a point.

Accordingly, if we wish to determine the direction of a tangent, we cannot in general hope to avoid detailed work with the secants getting closer and closer to the tangent.

Of course it must not be thought that this use of rulers is an accurate method.   If we need to establish some relationship in all seriousness in connexion with the direction of a curve, we should not dare to come forward with a result obtained by the mere turning of a ruler.   We cannot expect a precision instrument to be derived from drawing; this can come only from calculations.*

Let us begin with a definite example.   We shall try to follow the behaviour of the function given by the equation

$$Y = X^2$$

We already know that its picture is a parabola.   Let us try to decide with complete accuracy what the direction of its tangent is at the point whose X co-ordinate is 1.

At this point the Y co-ordinate is

$$Y = 1^2 = 1$$

---

* Those who are not curious to know about the differential and integral calculus and tend to get bored with fiddling details, may, as an exception, omit the remainder of this chapter as well as the next chapter.

so that our parabola passes through the point (1, 1). We are looking for the direction of the tangent drawn at the point (1, 1).

The picture, or graph, of the curve is an old friend:

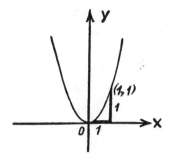

We know what to do: we must choose points on the curve which are progressively nearer and nearer to the point (1, 1). We must then draw secants through these points and the point (1, 1) respectively and determine the directions of these secants, for example in the form of quotients in terms of which we expressed the gradient of a railway line. Then we must find out to what direction these directions approximate as the secants approach the moment when they leave the curve.

We shall choose the neighbouring points in the following way: first we go to the right of the point (1, 1) by one unit, then by only one-tenth of a unit, then by one-hundredth, then by one-thousandth of a unit, and so on. So the X co-ordinates of the neighbouring points will be as follows:

$$1 + 1 = 2, \ 1 \cdot 1, \ 1 \cdot 01, \ 1 \cdot 001, \ \ldots$$

We shall need to calculate the Y co-ordinates of these points as well, and since $Y = X^2$, this will be done by squaring. This will be very easy, since 2^2 is of course 4, and we may perhaps still remember the second row in the Pascal Triangle (apart from the intervening zeros and decimal points, though actually we have had those too before now), so that

$$1 \cdot 1^2 = 1 \cdot 21, \ 1 \cdot 01^2 = 1 \cdot 0201, \ 1 \cdot 001^2 = 1 \cdot 002001, \ \ldots$$

There is just one more thing that needs to be said before we can go ahead, in order not to be worried by little details during more important considerations. We have done quite a lot of simplifying, so we now know that we can divide the numerator and the denominator of a fraction by one and the same number. If, for example, we simplify by 2 as follows:

$$\frac{6}{8} = \frac{3}{4}$$

we can also have, in reverse:

$$\frac{3}{4} = \frac{6}{8}$$

We can therefore also multiply the numerator and the denominator by 2, or by any other number. The form of the fraction thus becomes rather less simple, but we can make good use of this new knowledge if there are decimals in the fraction, for example if we are faced with a disagreeable kind of division such as

$$\frac{0 \cdot 21}{0 \cdot 1}$$

We know, of course, that in order to multiply a decimal by 10, we merely need to shift the decimal point one place to the right; it is unnecessary to write zeros in front of whole numbers, so multiplying the numerator and the denominator here by 10, we get

$$\frac{0 \cdot 2\ 1}{0 \cdot 1} = \frac{2 \cdot 1}{1} = 2 \cdot 1$$

In the same way the fraction

$$\frac{0 \cdot 0201}{0 \cdot 01}$$

becomes

$$\frac{0 \cdot 02\ 01}{0 \cdot 01} = \frac{2 \cdot 01}{1} = 2 \cdot 01$$

if we multiply the numerator and the denominator by 100, and so on.

Now we are in a position to begin. The first neighbouring number has 2 for its X co-ordinate, its Y co-ordinate is $2^2 = 4$. Let us draw the first secant through the points (1, 1) and (2, 4)

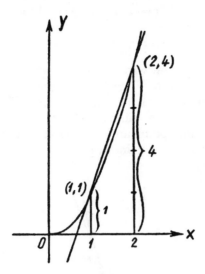

Now we must determine the direction of this secant. It is clear that we have moved one unit to the right from the point (1, 1), so this is the difference between the X co-ordinates. We have moved upwards 3 units, since this is the amount by which the Y co-ordinate of the second point has risen above our point. This is the difference between the Y co-ordinates. It can be seen on the diagram, in thick lines:

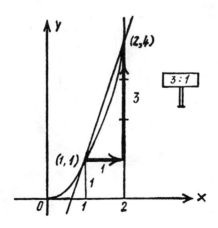

so the slope of the first secant is measured by 3 : 1, i.e.

$$\frac{3}{1} = 3 = 2 + 1$$

(there is a good reason why we should write it like this).

Now let us take the next neighbouring point.   Here $X = 1 \cdot 1$, and we have already calculated that $Y = 1 \cdot 1^2 = 1 \cdot 21$, so now we are dealing with the point $(1 \cdot 1, 1 \cdot 21)$.   If we try to draw a secant through this point and our point (the rise is again indicated by means of thick lines), parts of the figure are so small, they almost coalesce:

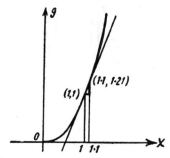

Let us put the relevant bit of the drawing under a magnifying glass:

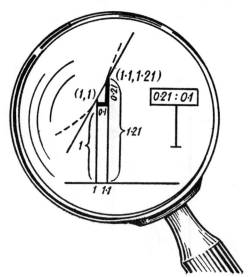

How far have we moved to the right? By 0·1, since this is the difference between the X co-ordinates. By how much has the Y co-ordinate of the second point risen above our point? By as much as the difference between the two Y co-ordinates, i.e. by

$$1·21 - 1 = 0·21$$

units. So the second secant has a slope which is measured by

$$0·21 : 0·1$$

i.e.
$$\frac{0·21}{0·1}$$

which, as we have seen, is the same as

$$2·1 = 2 + \frac{1}{10}$$

If we go on to the next neighbouring point for which $X = 0·01$ and, as we have already found, $Y = 1·01^2 = 1·0201$, we should need a much more powerful magnifying glass. But perhaps we may now dispense with drawing, since, as we have noticed, we always have to divide the difference of the Y co-ordinates by the difference of the X co-ordinates. The Y co-ordinates at this point and at the original point differ by

$$1·0201 - 1 = 0·0201$$

and the X co-ordinates of these points differ by

$$1·01 - 1 = 0·01$$

so the slope of the third secant is measured by

$$0·0201 : 0·01$$

i.e. by
$$\frac{0·0201}{0·01}$$

which we have already seen is equal to

$$2·01 = 2 + \frac{1}{100}$$

We can go on in this way with the result that the quotients of the Y differences and the X differences (for short, the 'difference quotients') of the secants, as they come closer and closer to the starting point on the curve, have the values

$$2 + 1, \ 2 + \frac{1}{10}, \ 2 + \frac{1}{100}, \ 2 + \frac{1}{1000}, \ \cdots$$

respectively.

We know that the sequence

$$1, \frac{1}{10}, \frac{1}{100}, \frac{1}{1000}, \ldots$$

converges to zero with the precision of the 'chocolate example'.*
The number to which the above slopes approximate more and
more is therefore exactly

$$2$$

with one hundred per cent precision.   Where the secant just
leaves the curve, it becomes the tangent, so that the slope of
the tangent drawn to the parabola at its point (1, 1) is 2, i.e. $\frac{2}{1}$.
On this basis we can draw the tangent:

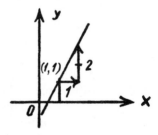

and, if we construct the parabola near it with the aid of a number of intermediate values, we shall have the definite feeling
that this straight line touches the parabola:

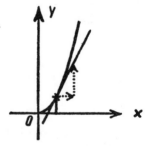

It appears that, while drawing our pictures, a completely
accurate method of computing the direction of the tangent has
fallen into our hands as a by-product; we must choose another
point on the curve in the vicinity of our point, then we must
divide the difference of the Y co-ordinates of these points by the

* See pages 105–106.

difference of the corresponding X co-ordinates, and find out to what the quotients so obtained approximate, as the neighbouring point approaches our point.

The quotients of such differences are called the difference quotients and the definite value to which the difference quotients approximate is called the differential quotient, or differential coefficient. Differentiation is therefore what I said it would be: a precise process for determining the tangents of a smooth curve, as well as for examining the behaviour of the entire curve.

The procedure can be applied at other points too; if the curve is smooth, it will have a definite direction at every one of its points. At the point (2, 4) we feel that the parabola is steeper, and if we calculate the difference quotients corresponding to the points

$$X = 2 + 1,\ 2\cdot1,\ 2\cdot01,\ 2\cdot001,\ \ldots$$

we obtain

$$4 + 1,\ 4 + \frac{1}{10},\ 4 + \frac{1}{100},\ 4 + \frac{1}{1000},\ \cdots$$

respectively, and 4 is the number, with absolute accuracy, to which these approximate more and more. Therefore the slope of the tangent at the point (2, 4) is $4 = \frac{4}{1}$, and this is in fact more than the slope of the tangent drawn at the point (1, 1), which was $2 = \frac{2}{1}$.

It can be shown in just the same way that the slope of the tangent corresponding to the point $X = 3$ is 6, the slope of the tangent corresponding to the point $X = 4$ is 8; in general the value of the slope at every point of the parabola is twice the value of the X co-ordinate of the point in question. This is expressed by saying that the differential coefficient of the function

$$Y = X^2$$

for any arbitrary value of X is

$$\underline{\underline{2X}}$$

And this does in fact give us the clue to the behaviour of the whole parabola.

To begin with something definite, let us note from the equation of the function that the curve passes through the zero point, since if $X = 0$, then $Y = X^2 = 0^2 = 0$. The rest of the information is provided by the differential coefficient.

Let X for example be a negative number. Its double $2X$ is then likewise negative, so that the slope of the tangent is negative; at a point like this the tangent drops downwards and with it the curve that clings to it. If, on the other hand, X is positive, then its double is likewise positive and the curve rises upwards at such a point. If $X = 0$ then its double $2X$ is also 0, so that at the zero point the slope of its tangent is zero. A zero slope is of course not a slope, i.e. it is a horizontal path; the path here is the X axis itself. As the absolute value of X increases, its double increases with it more and more and with it the steepness of the tangent.

To sum up, we obtain the following picture of the curve: to the left of zero point the slope falls away, at the zero point it becomes horizontal for a moment and clings to the X axis, and from here onwards it goes on rising. It follows that its lowest point is at the zero point. As we go farther away from the zero point, whether to the right or to the left, both sides of the curve will get steeper. Of course we already know all this about the parabola, but in the case of a lesser-known function the differential coefficient would have provided all this information.

The knowledge of the differential coefficient may increase the exactness of our existing knowledge about the parabola. We saw when we drew the first few charts that the picture of the multiplication function

$$2X$$

was a straight line (this was to be expected, since it is linear); therefore this function increases at an even rate. It follows that the steepness of each side of the parabola increases, not capriciously or with greater intensity, but quite gradually.

It may happen that the parabola is shifted from its usual place

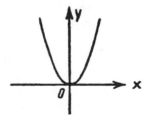

to some other positions, for example into positions like these:

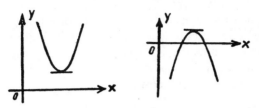

and it then becomes a problem to decide where its lowest or highest point might be. The differential coefficient gives an immediate answer to this question, since the tangent of the parabola will still be horizontal at such points.

The search for such lowest or highest points, or, in the language of Function Theory, the determination of the maxima and the minima of functions, can have a great number of practical applications.

For example, if we want to make a box out of a square piece of material by cutting out small squares from the four corners and folding the remaining pieces upwards,

the question arises what size squares must we cut out in order to obtain a box of maximum volume?

We do not know the length of the side of the small square, so let us call it X. It is quite an easy matter to determine in what way the volume of the box depends on the choice of X. It is obvious that if X is small, i.e. if we cut out only a little, we shall get a low, wide box; if we cut away larger squares, then we shall get a smaller base and the box will be taller but narrower. So we must not make X either too small or too big, the right value must lie somewhere in between. The differential co-efficient establishes with perfect accuracy that we shall obtain a box of maximum volume if the side of the small square is exactly $\frac{1}{6}$ of the side of the large square.

A heavy stone is flying through the air; we know the greatest height it will reach, since the differential coefficient tells us exactly the highest point of a projectile.

The applications are too numerous to be counted.

Let us examine a case where the curve of the function is not such an old friend as the parabola. In exactly the same way, by examining the difference quotients, it can be established that the function given by the equation

$$Y = X^3$$

has a tangent at each point whose slope is exactly 3 times the square of the X co-ordinate of the point, i.e. the differential coefficient of the function is

$$3X^2$$

What information can we gather from this?

In order to have a concrete starting point, we observe from the equation of the function itself that

$$\text{if } X = 0 \text{ then } Y = 0^3 = 0$$

and so the curve passes through the zero point.

Now let us see what the differential coefficient has to say.

The first thing that strikes us is that X occurs in it squared (its own picture is a parabola). We can draw two conclusions from this: one is that in the case of the curve of $Y = X^3$ there can be no question of a uniform growth of its steepness; this steepness becomes more and more precipitous as we move away from the zero point. The other conclusion is that whether we are looking at a positive or at a negative X co-ordinate, X^2 is always going to be positive, so that the tangent, and, with it, the curve, must be a rising curve both on the left and on the right of the zero point. Since the curve passes through the zero point, the only way it can be a rising one before the zero point is by remaining below zero, below the X axis. After the zero point the curve rises above this height, so it must cut the X axis at the zero point. But

$$\text{if } X = 0, \text{ then } 3X^2 = 3 \times 0^2 = 3 \times 0 = 0$$

so the slope of the tangent at the zero point is itself zero, and the tangent here must be horizontal. The horizontal line passing through the zero point is, of course, again the X axis. The X axis therefore touches the curve and at the same time cuts it at the zero point. Approaching this point from the left the slope becomes gentler, it has a moment's rest at the actual point, and then again gathers fresh strength and begins to rise, becoming steep very rapidly.

On the basis of the above the picture we have formed of the curve is something like this:

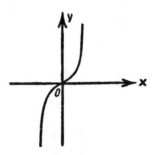

Now let us represent the function

$$Y = X^3$$

on a graph.

If $X =$ 0 then $Y =$ $0^3 =$ 0
„ $X =$ 1 then $Y =$ $1^3 =$ 1
„ $X =$ 2 then $Y =$ $2^3 =$ 8
„ $X = -1$ then $Y = (-1)^3 = -1$
„ $X = -2$ then $Y = (-2)^3_4 = -8$

and, taking a few intermediate values:

If $X = \dfrac{1}{2}$ then $Y =$ $\left(\dfrac{1}{2}\right)^3 = \dfrac{1}{2} \times \dfrac{1}{2} \times \dfrac{1}{2} = \dfrac{1}{8}$

„ $X = -\dfrac{1}{2}$ then $Y = \left(-\dfrac{1}{2}\right)^3 = -\dfrac{1}{8}$

„ $X = \dfrac{1}{4}$ then $Y =$ $\left(\dfrac{1}{4}\right)^3 = \dfrac{1}{4} \times \dfrac{1}{4} \times \dfrac{1}{4} = \dfrac{1}{64}$

For the representation of $\frac{1}{64}$ a pencilled point would come too high; it appears from our drawing that the curve already clings to the X axis at this point (a more thorough examination of the differential coefficient would have predicted this closer form of clinging). We have seen that at the points

$$0, \quad \dfrac{1}{2}, \quad 1, \quad 2$$

we must measure

$$0, \quad \frac{1}{8}, \quad 1, \quad 8$$

units upwards, and at the points

$$-\frac{1}{2}, \quad -1, \quad -2$$

we must measure

$$-\frac{1}{8}, \quad -1, \quad -8$$

units downwards:

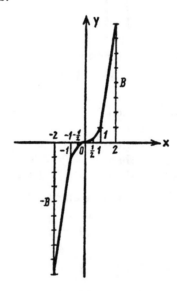

This is in fact the picture that the differential coefficient predicted. There cannot be any freak variation at the intermediate points either, as the differential coefficient would have predicted these too. It goes without saying that it does not only predict the approximate picture, but determines the direction of the curve with perfect accuracy at every one of its points.

It is small wonder that mathematicians have taken the trouble to determine the differential coefficients of all the functions they are likely to come across, and that they have worked with these such a lot that they know them all by heart.

Whenever a physicist fetches a function out of his mathematical store-cupboard, he will always find with it, as one of the most important instructions for use, the differential coefficient of the function, thoughtfully provided by mathematicians.

17. 'Many small make a great'

WE have done so many multiplications in our lives that we
know our tables completely by heart, and so in the case of the
inverse operation we immediately recognize 5 as the number
which if we multiply by 4 will give us 20. Mathematicians
know the differential coefficients of all the usual functions by
heart, and recognize them when they see them. If somebody
should mention the function $2X$, even we should feel that it
is somehow familiar. Where did we come across it? Of
course, it was the differential coefficient of the function X^2. So
we can speak about the inversion of an operation here too.
Given a function, we might ask if there is another function of
which our function is the differential coefficient, and, if there is
one, what function is it? If there is one, it is called the integral,
for example the integral of $2X$ is the function X^2. There are
tricks here, too, which facilitate the finding of the function we
are looking for, in the same sort of way as in the case of equations, if we do not recognize a function straight away as a differential coefficient. Let us take, for example, X^2 as our given
function. This may remind us to some extent of the function
$3X^2$ which we already know is the differential coefficient of the
function given by the equation $Y = X^3$. Our X^2, whatever X
is, is just one-third of $3X^2$. Perhaps it will be the differential
coefficient of one-third of X^3, i.e. of the function

$$Y = \frac{X^3}{3}$$

It is quite easy to show that this is in fact the case.

In most cases, unfortunately, tricks are of no avail. There is
a need for more general methods. And there is something a
little wrong with the above guessing method. The differential
coefficient does not tell us that for example the curve of the
function $Y = X^2$ passes through the 0 point. We had to read
this off the equation of the function itself at the time. How can
we imagine, then, that the differential coefficient is sufficient for
the complete determination of the curve?

196

In fact it is not quite enough, as we can see in a moment. Let us shift our parabola upwards by one unit:

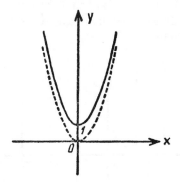

It is obvious that the shape of the curve is not affected by a mere shift. Its steepness is the same at every one of its points, so that the differential coefficient is still the same. On the other hand, the equation of the curve must have been altered, since the Y co-ordinate of every point has become 1 unit more than it was before. Therefore any Y which was previously X^2 has now become $X^2 + 1$, so the equation of the shifted parabola is

$$Y = X^2 + 1$$

From the variation of the direction alone, i.e. from the differential coefficient, it is not possible to find out whether we meant this function or the other one, or any of the innumerable parabolas which we might have obtained by shifting our original parabola up and down. To this extent our problem remains indeterminate.

If, on the other hand, we give one single point of the required curve, then the problem becomes determinate. If, for example, we say that as an 'initial value' we require the curve to pass through the zero point, then from our differential coefficient we can get only our original parabola. This will be made clear from what follows.

We shall show the general method, using the parabola. Let us suppose that we do not recognize the integral of the function $2X$. Let us try to determine the curve about which all we know is that it *passes through the 0 point*, and that the slope of its tangent at any point is $2X$.

Let us start with a drawing, though here again our eventual aim will be to find a precise method.

Let us subdivide the X axis at first into unit intervals, and at the points of subdivision let us draw vertical lines for the unknown Y co-ordinates.

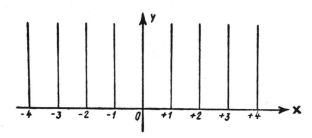

It is only at the zero point that we know that Y is also zero. Let us start drawing the curve here, only approximately of course.

The basic idea behind the drawing must be that the tangent clings to the curve for a little while, so that for a short distance from the point of contact it may still be used as a reasonable substitute for the curve. We shall assume such a short distance to be the distance between two consecutive verticals on our drawing. At first we draw the tangent corresponding to the 0 point, and we may assume that this tangent will represent our curve up to $+1$ on the right and up to -1 on the left. The points reached in this way we can regard as the points on the curve corresponding to $X = +1$ or to $X = -1$ respectively, and starting from these we can draw the corresponding tangents up as far as the next vertical lines. The points we thus obtain we can consider as the points on the curve corresponding to $X = +2$ and $X = -2$ respectively. We can then draw the tangents at these points up as far as the next vertical lines and so on. The tangents must, of course, be drawn in accordance with the given slope. At the 0 point this is

$$2X = 2 \times 0 = 0$$

at the point $X = 1$

$$2X = 2 \times 1 = 2$$

and as we know that the product function $2X$ increases uniformly at successive points after $X = 1$ the slopes will be 4, 6,

8, . . .; similarly from 0 towards the left they will be −2, −4, −6, . . . Accordingly the slope of the tangent at the points

$$\frac{0 \mid 1 \mid 2 \mid -1 \mid -2}{0 \quad 2 \quad 4 \quad -2 \quad -4}$$ will be respectively

Of course we know that a slope of 2, i.e. of $\frac{2}{1}$, means that, if we go one unit to the right, we must rise by 2 units, and similarly that a slope of −2 means that, if we go one unit to the left, we rise by 2 units. So, for example, we measure the same amount upwards at the points $X = +1$ and $X = -1$, and it follows that the drawing is symmetrical. It is enough to draw the right-hand side of the drawing accurately, since we can then copy it for the left-hand side.

Now we can start drawing. At the 0 point the value 0 of the slope tells us that a zero slope means a horizontal path; we proceed horizontally as far as the point 1, then we proceed from there with a slope of $2 = \frac{2}{1}$ as far as the next vertical, from here our path will have a slope of $4 = \frac{4}{1}$ to carry us farther:

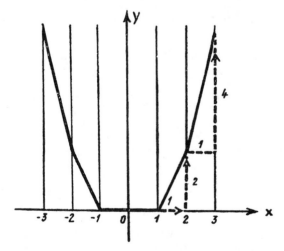

We obtain a somewhat rough figure of the parabola.

Let us check the accuracy of our result by calculation. Let us restrict ourselves to the point $X = 3$ and calculate the Y co-ordinate of the curve $Y = X^2$ at this point.

$$\text{If } X = 3 \text{ then } Y = 3^2 = 9$$

Of course we are not yet supposed to know that we are dealing with the function $Y = X^2$, but since we nevertheless do secretly know it, we can use it as our measure. We shall see to what extent our bent line has a Y co-ordinate different from 9 at $X = 3$.

We see from our drawing that we have reached the Y co-ordinate in question by going from the zero point and adding up all the rises in between the vertical lines, so that for us

$$Y = 0 + 2 + 4 = 6 = 9 - 3$$

and 3 is quite a big difference. Let us take more points of sub-division by drawing vertical lines at intervals of $\frac{1}{2}$ a unit:

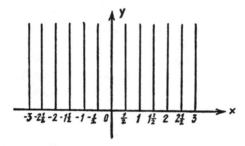

The slope of the tangent at the 0 point is still zero (the tangent is horizontal). At the point $X = \frac{1}{2}$

$$2X = 2 \times \frac{1}{2} = 1$$

and in the equal intervals the slope increases uniformly, so that it will be always 1 more at the successive points of subdivision.

The slope of the tangent at the points

$0|\frac{1}{2}|1|1\frac{1}{2}|2|2\frac{1}{2}|-\frac{1}{2}|-1|-1\frac{1}{2}|-2|-2\frac{1}{2}$ will be respectively
0 1 2 3 4 5 -1 -2 -3 -4 -5

There is just one thing we must be careful about before we begin to draw this. At the 0 point the slope is 0 and from here we must proceed horizontally as far as the point $\frac{1}{2}$. But at the point $\frac{1}{2}$ the slope is 1, i.e. $\frac{1}{1}$, so that from here we ought to go 1 unit to the right and 1 unit upwards. Now we did not draw our vertical lines at every $\frac{1}{2}$-unit with the idea of proceeding by whole units to the right as before. We must realize that if a railway line has slope $\frac{1}{1}$, then walking alongside it on a horizontal path for 1 yard, the line would rise 1 yard, and that if we walked $\frac{1}{2}$ a yard alongside it on the horizontal path the line would rise $\frac{1}{2}$ a yard:

In the same way, if the railway line has slope 2 $= \frac{2}{1}$, then if we walk $\frac{1}{2}$ a yard beside it horizontally instead of 1 yard, then it will rise not by 2 yards but by 1 yard. Therefore, if we proceed by $\frac{1}{2}$-units, we have to take $\frac{1}{2}$ of the slopes just calculated as representing the amounts of actual rise. For example, starting from 0 and proceeding towards the right at the points

| 0 | $\frac{1}{2}$ | 1 | $1\frac{1}{2}$ | 2 | $2\frac{1}{2}$ |

we do not rise

| 0 | 1 | 2 | 3 | 4 | 5 |

but

0 $\frac{1}{2} \times 1 = \frac{1}{2}$ $\frac{1}{2} \times 2 = 1$ $\frac{1}{2} \times 3 = 1\frac{1}{2}$ $\frac{1}{2} \times 4 = 2$ $1 \times 5 = 2\frac{1}{2}$

Now we can prepare our drawing without any further difficulty.

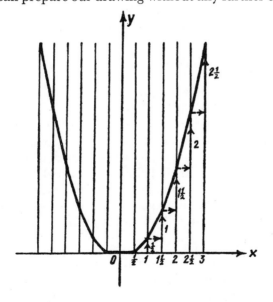

This looks as though it is going to smooth out into a parabola; it merely exaggerates the clinging to the X axis.

Let us again calculate the Y co-ordinate corresponding to the point $X = 3$.

This time it is not added up from the slopes, but from half of these slopes. It is better not to write these in their computed forms

$$0 \qquad \tfrac{1}{2} \qquad 1 \qquad 1\tfrac{1}{2} \qquad 2 \qquad 2\tfrac{1}{2}$$

but rather in the forms

$$\tfrac{1}{2} \times 0 \quad \tfrac{1}{2} \times 1 \quad \tfrac{1}{2} \times 2 \quad \tfrac{1}{2} \times 3 \quad \tfrac{1}{2} \times 4 \quad \tfrac{1}{2} \times 5$$

so that we have

$$Y = \tfrac{1}{2} \times 0 + \tfrac{1}{2} \times 1 + \tfrac{1}{2} \times 2 + \tfrac{1}{2} \times 3 + \tfrac{1}{2} \times 4 + \tfrac{1}{2} \times 5$$

Since $\tfrac{1}{2} \times 0 = 0$, we can leave this one out.

If we have to take half of every term, it is simpler to add the terms first and then take half of the result

$$Y = (1 + 2 + 3 + 4 + 5) \times \tfrac{1}{2}$$

In this way we need only to add whole numbers in the brackets. Even these can be added up more neatly by the method of my pupil Susie. We can take their 'middle', i.e. 3, 5 times, this is 15; this must now be taken $\tfrac{1}{2}$ times, that will be $\tfrac{15}{2}$. If we added 3 more to 15, we should get the number 18, which is divisible by 9, so that finally we have

$$Y = \frac{15}{2} = \frac{18}{2} - \frac{3}{2} = 9 - \frac{3}{2}$$

The corresponding Y co-ordinate of the previous curve differed from 9 by 3; this one only differs by $\tfrac{3}{2}$.

While our bent line gradually gets smoothed out (the process can give us only approximate results on account of the imperfect tools used) we obtain as a by-product, a computational procedure, which can be indefinitely perfected, for calculating the Y co-ordinate corresponding to the point $X = 3$. It should be obvious that if we go on to subdivisions of $\tfrac{1}{4}$ of a unit, the slope at the 0 point will still be 0, at the point $X = \tfrac{1}{4}$

$$2X = 2 \times \frac{1}{4} = \frac{2}{4} \text{ or, simplifying: } \frac{1}{2}$$

and so the slope will increase always by $\tfrac{1}{2}$ in the equal intervals, so the slope of the tangent will be, starting from the point 0,

at the points

$$0, \quad \frac{1}{4} \middle| \frac{2}{4} = \frac{1}{2} \middle| \frac{3}{4} \middle| 1, \middle| 1\frac{1}{4} \middle| 1\frac{1}{2} \middle| 1\frac{3}{4} \middle| 2, \middle| 2\frac{1}{4} \middle| 2\frac{1}{2} \middle| 2\frac{3}{4}$$

$$0, \quad \frac{1}{2}, \quad \frac{2}{2}, \quad \frac{3}{2} \frac{4}{2} \frac{5}{2} \frac{6}{2} \frac{7}{2} \frac{8}{2} \frac{9}{2} \frac{10}{2} \frac{11}{2}$$

In this case we want to proceed to the right in $\frac{1}{4}$ units, so that only a quarter of these figures will be the actual amounts of rise at the corresponding points, for if we walk only one-quarter of the distance horizontally alongside a sloping railway line, then the railway line also only rises one-quarter of the amount. Our Y will therefore be compounded out of these quarters, until we reach the point $X = 3$, i.e.

$$Y = \frac{1}{4} \times \frac{1}{2} + \frac{1}{4} \times \frac{2}{2} + \frac{1}{4} \times \frac{3}{2} + \frac{1}{4} \times \frac{4}{2} + \frac{1}{4} \times \frac{5}{2} +$$

$$+ \frac{1}{4} \times \frac{6}{2} + \frac{1}{4} \times \frac{7}{2} + \frac{1}{4} \times \frac{8}{2} + \frac{1}{4} \times \frac{9}{2} + \frac{1}{4} \times \frac{10}{2} + \frac{1}{4} \times \frac{11}{2}$$

As $\frac{1}{4} \times 0 = 0$ it can be left out.

Here every term has to be taken $\frac{1}{4}$ times, in other words we must divide by 4. Apart from this the denominator of each fraction indicates a further division by 2. We know that, if we divide something by 4 and then by 2, we shall get the same amount as if we divide straight away by $4 \times 2 = 8$. Apart from this, the terms to be divided can be added first and the result then divided by 8. We therefore have

$$Y = (1 + 2 + 3 + 4 + 5 + 6 + 7 + 8 + 9 + 10 + 11) \times \frac{1}{8}$$

With so many terms, we are really lucky in having Susie's method at our disposal. We need only to take the middle one of the terms, i.e. 6, and multiply this by 11, this will be 66; but we must divide by 8, that will be $\frac{66}{8}$. We could add another 6 to 66 in order to get 72, which is a number divisible by 9, so that finally

$$Y = \frac{66}{8} = \frac{72}{8} - \frac{6}{8} = 9 - \frac{6}{8}$$

but $\frac{6}{8}$ can be simplified by 2, so that

$$Y = 9 - \frac{3}{4}$$

In this refinement only $\frac{3}{4}$ are missing from the 9.

This result was obtained without any recourse to drawing, but we still had to think what would happen if we did do the drawing. We can continue the process without even thinking about any drawing. The next step would be the subdivision of the distance between 0 and 3 into $\frac{1}{8}$ths of a unit. At the successive points of subdivision the slope would increase by steps of amount

$$2X = 2 \times \frac{1}{8} = \frac{2}{8} = \frac{1}{4}$$

so at these points the slopes would be

$$0, \ \tfrac{1}{4}, \ \tfrac{2}{4}, \ \tfrac{3}{4}, \ \tfrac{4}{4}, \ \tfrac{5}{4}, \ \ldots$$

We should have to multiply these numbers in turn by the length $\frac{1}{8}$ of the intervals and then we should have to add up these numbers as far as the point $X = 3$. The result would be

$$Y = 9 - \frac{3}{8}$$

It can easily be seen that this can be continued indefinitely. The sequence

$$3, \ \tfrac{3}{2}, \ \tfrac{3}{4}, \ \tfrac{3}{8}, \ \ldots$$

converges to 0 (if we divided 3 cakes among more and more people, each person would get a more and more negligible amount), so that 9 is the number, *with perfect accuracy*, which is being approximated by the Y co-ordinates corresponding to $X = 3$ on our curves as they get more and more smoothed out; 9, in other symbols 3^2, i.e. the value of the function $Y = X^2$ at the point $X = 3$.

It can be proved in the same way that the Y co-ordinates of our curves at $X = 1$ converge to $1 = 1^2$, at $X = 2$ to $4 = 2^2$, at $X = 4$ to $16 = 4^2$, in general at any point the Y co-ordinates converge to the square of the X co-ordinate in question, i.e. to X^2, so that our bent lines will finally get smoothed out into the parabola

$$Y = X^2$$

Or in the language of functions: if just one initial value is given, it is possible to reconstruct from it the function

$$2X$$

namely that function of which it is the differential coefficient.

During our labours we actually came across the required precise method for doing this. The X axis must be subdivided into intervals from the given point as far as the point to be examined (in our case from 0 to 3), the length of the interval must each time be multiplied by the value of the function at the point of subdivision, and all these products must be added up. In this way we get 'approximate integral sums'. If we take the points of subdivision more and more densely over the relevant interval, these sums converge to the value of the integral at the point examined. It must be admitted that on the whole this is rather an awkward process. But, as we have already seen, inverse operations tend to be bitter operations.

It is actually possible to represent the approximate sums by means of areas. Each term of every approximate sum is a product: we multiply the length of the interval by the value of the function at some point or other. But we know by now that we can represent a product by means of the area of a rectangle whose adjacent sides are the lengths of the two factors. In this way every term of the approximate sum gives us a rectangle. We can represent the whole sum by simply putting all these rectangles next to one another.

Let us have a try. Our first sum was

$$0 + 2 + 4$$

in which we cannot see the products, since in this case the length of the intervals was one unit; so let us write our sum in the following form:

$$1 \times 0 + 1 \times 2 + 1 \times 4$$

Now we can represent it:

(1 $\times$ 0 can be considered as a rectangle reaching from 0 to 1 and of
0 height; this of course is just a
horizontal segment.)

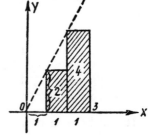

Our second approximate sum was this:

$$\tfrac{1}{2} \times 0 + \tfrac{1}{2} \times 1 + \tfrac{1}{2} \times 2 + \tfrac{1}{2} \times 3 + \tfrac{1}{2} \times 4 + \tfrac{1}{2} \times 5$$

The picture of this will be

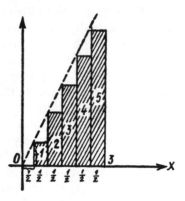

Our third approximate sum had 12 terms:

$$\frac{1}{4} \times 0 + \frac{1}{4} \times \frac{1}{2} + \frac{1}{4} \times \frac{2}{2} + \frac{1}{4} \times \frac{3}{2} + \frac{1}{4} \times \frac{4}{2} + \frac{1}{4} \times \frac{5}{2} +$$

$$+ \frac{1}{4} \times \frac{6}{2} + \frac{1}{4} \times \frac{7}{2} + \frac{1}{4} \times \frac{8}{2} + \frac{1}{4} \times \frac{9}{2} + \frac{1}{4} \times \frac{10}{2} + \frac{1}{4} \times \frac{11}{2}$$

This can easily be represented by means of $\tfrac{1}{2}$-units. As there really is no room now for writing out the numbers fully, the drawing will have to suffice:

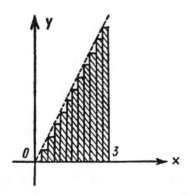

It will be seen that these 'staircases' approximate more and more to the area of a right-angled triangle. I am thinking of the triangle that lies below the dotted line in all the figures. Perhaps the reader will have noticed that this straight line is the same in all the figures. From the first figure we can read off quite easily that its slope is

$$2 : 1$$

and we can check on the other figures that it is just the same in each. Only a short time ago I suggested that the reader might be able to recognize it. The straight line passing through the 0 point whose slope is 2 : 1 is the straight line whose equation is

$$Y = 2X$$

But this is just the function that we are given! This straight line is the exact picture of our function. So the approximate sums actually approximate more and more accurately to the area below the picture of the curve. What a pity that we did not know this before, since it is very easy to calculate the area of a right-angled triangle; we have only to multiply the two sides adjacent to the right angle and take half of this. The horizontal adjacent side is the part stretching as far as $X = 3$, i.e. its length is 3 units. Let us calculate the vertical one:

$$\text{if } X = 3, \text{ then } Y = 2X = 2 \times 3 = 6$$

so the other side adjacent to the right angle is 6 units:

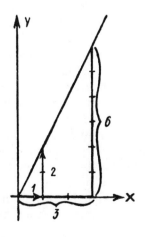

The area of the triangle is therefore

$$\frac{3 \times 6}{2} = \frac{18}{2} = 9 \text{ units}$$

and this agrees with the result obtained much more laboriously just now.

In this way the calculation of areas can help us to calculate integrals. This, in fact, is not just a matter of chance. As long as the function we are dealing with is not too wild, like the Dirichlet function which goes on jumping about between 0 and 1 all the time (in which case the approximate integral sums have no intention whatever of converging), i.e. in the case of the more normal type of function, the approximate sums can always be represented by means of areas of such 'staircases':

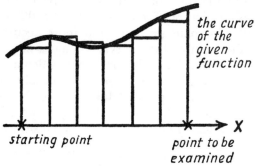

the curve
of the
given
function

starting point

point to be
examined

X

and these approximate to the area lying below the curve corresponding to the function with the accuracy of the 'chocolate example', from the initial point to the point under examination, as long as we make the subdivisions indefinitely more and more dense. In other words: the area below the curve and the integral are one and the same concept, only expressed differently.

The roles are often reversed and the calculation of areas can often be made much easier by the calculation of integrals.

We can calculate the area of a right-angled triangle, and we know that other triangles can be split into right-angled triangles and all polygons can be split into triangles. We see that the calculation of the areas of figures bounded by straight lines is not a problem. We have also somehow reconciled ourselves to calculating the area of a circle by means of stuffing a large number of very thin triangles inside it. But what about calculating in general the area bounded by a curve?

Such areas can be cut up by means of straight lines, and each piece can then be fitted with its straight side along the X axis:

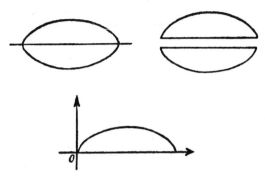

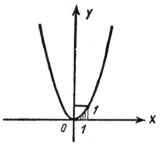

We can then calculate the area of each piece separately. The calculation of the area lying under such a curve is a problem in the integral calculus. It may happen that we hit upon an integral that can readily be guessed, and in such cases we can say in a few moments what the corresponding area must be.

For example we have found out that the integral of X^2 is the function $Y = \dfrac{X^3}{3}$ or rather, more accurately, that, out of all the possible functions, this is the one that passes through the 0 point, because

$$\text{if } X = 0 \text{ then } \frac{X^3}{3} = \frac{0^3}{3} = 0$$

From this we can calculate in a few moments the area under the parabola

$$Y = X^2$$

For example, as far as the point $X = 1$ this is equal to the value of the integral at $X = 1$, i.e.

$$\frac{1^3}{3} = \frac{1}{3} \text{ units of area}$$

The shaded area, which clearly is only a part of a unit square, is exactly $\frac{1}{3}$ of this unit square.

Perhaps it may not be considered very interesting to know the area *outside* a parabola. This may be so, but from our result we can calculate the area between the two branches of our parabola to any desired height. For example, if the third of the above unit square is outside, $\frac{2}{3}$ of it will be inside, and if we add to this its mirror-image on the left, then the shaded area in the figure below will have an area of

$$2 \times \frac{2}{3} = \frac{4}{3} = \frac{3}{3} + \frac{1}{3} = 1\frac{1}{3} \text{ units}$$

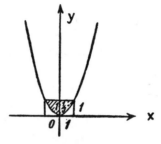

I should again like to draw the reader's attention to the number of little rectangles by means of which we have approximated the area:

As the subdivision gets more dense, the rectangles become thinner. The area of each rectangle necessarily converges to zero, in the sense of the cake, divided into many pieces, already mentioned so frequently. But these little slices, although becoming thinner and getting nearer and nearer to zero thickness, together nevertheless approximate to a definite area different from zero, and this area need not even be small. The area of our triangle discussed before was in fact 9 units. This is not surprising since, as the rectangles get thinner, they get more and more numerous, and a lot of small things finally add up to a big thing. The depositing of almost invisible layers of sand in time buries even the largest pyramids. A lot of little people think something, and the world suddenly takes an important turn. A lot of small effects get 'integrated'.

PART III

THE SELF-CRITIQUE OF PURE REASON

18. And still there are different kinds of Mathematics

THERE is hardly a well-known mathematician to whom it has
not happened that a mysterious stranger has handed him over a
greatly treasured manuscript, sometimes bulky, sometimes
quite short, in which the squaring of the circle is 'accomplished'.
Let us see what this really means.

If somebody says: 'I knew the two sides of a right-angled
triangle adjacent to the right angle, and I was able to construct the whole triangle from these data', the question
immediately arises: 'What tools did you use?' Suppose that
he used a right-angled triangle made of wood, obtainable at a
shop:

in other words he drew his pencil along the sides of this object.
Of course it would not be wise to rely on the accuracy of such
articles. 'Turn the wooden triangle round and place it next
to the right angle just drawn with its aid, and draw some
straight lines with it.' In most cases the result will be something like this

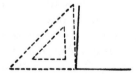

so that the wooden triangle is not exactly right-angled.

The ancient Greeks took great care to choose the tools which they were prepared to use in their constructions.   Rulers were to be used only for drawing single straight lines along them (they were not allowed to use them for drawing right angles); of course even that is a compromise, since it often happens that the edge of a ruler is not really quite straight. If we want to draw a circle, we can do this with a much more accurate instrument; there is no need to draw round any ready-made wooden circle, for with a pair of compasses we can form the circle ourselves.   If the two parts of the compasses are not hinged too loosely, we can fix the pointed end of one part by sticking it into the paper at one fixed point, the pencil end of the other half will in fact move at a constant distance from our fixed point, and so will describe a real circle.

moving point   fixed point

The ancient Greeks did not allow any other instruments for their geometrical constructions.   The constructions were more reliable, the more they depended on the compasses alone, and the less often it was necessary to have recourse to the ruler. After several centuries it came about that there was no need for any ruler at all.   All the constructions that could be done by means of ruler and compasses alone, could be done by using the compasses only.   Of course it is not possible to draw a straight line with compasses, but, for example, a square can be represented by its four vertices:

and we can imagine the figure quite well even from points like these.

But let us keep to ruler and compasses. The question arises quite naturally: what constructions may be done by means of these two tools alone?

The problem of squaring the circle belongs to this category of problems. Given a circle, it is required to construct a square whose area is exactly equal to the area of the given circle.

We already know that it is possible to determine the area of the circle with perfect accuracy, by means of other areas bounded by straight lines, getting nearer and nearer to the circle. For example if we have drawn a circle with a unit radius, we obtain a definite irrational number for the measure of its area; this number begins like

$$3 \cdot 14 \ldots$$

and the calculation of this number to any desired degree of accuracy can be carried out. This irrational number plays such an important part in Mathematics that it has received a special name. This is the number

$$\pi$$

well known from our school days.

If we really know the area of the circle of unit radius as accurately as all this, we can of course say straight away which is the square whose area is this amount. We calculate the area of a square by squaring the length of one of its sides. There is of course a number whose square is π; this is what we denote by $\sqrt{\pi}$. So the square whose side is $\sqrt{\pi}$ solves the problem.

But the problem was not whether there was such a square, but whether it could be constructed accurately by using ruler and compasses only.

The fact that $\sqrt{\pi}$ is irrational need not necessarily hinder the construction, since we have already drawn squares whose sides were $\sqrt{2}$. The reader will perhaps recall doubling the size of the fishpond. The argument sketched on that occasion could very easily be transformed into an accurate construction. Would it not be possible somehow or other to construct $\sqrt{\pi}$ by means of ruler and compasses alone?

Many people attacked this problem for centuries without success. Finally the translation of the problem into the language of Algebra led to its solution.

What can we draw with rulers and compasses? Straight lines and circles. We already know that in the language of Algebra straight lines mean linear equations and circles certain types of quadratic equations. Anything that can be constructed by means of rulers and compasses will have to appear as a common solution to such equations.

Now mathematicians have succeeded in proving that $\sqrt{\pi}$ (or even π) cannot be a solution of any such equations or, come to that, of any equations at all, of however high a degree, unless π is somehow smuggled into the equation first (for example from the equation

$$X - \pi = 0$$

if we take π over to the right as a term to be added, we have

$$X = \pi)$$

We say that π is not an 'algebraic' number, it is a 'transcendental' number.

Looked at in this light, the squaring of the circle is an insoluble problem. Mathematics has again succeeded brilliantly in demonstrating its own inefficacy in the solution of a problem where the methods of solution are clearly circumscribed.

Apart from the discovery of the existence of 'transcendental' numbers, which cannot occur among the solutions of any kind of algebraic equation (it can be shown that $e = 2 \cdot 71 \ldots$, the base of natural logarithms is also such a number, moreover that the vast majority of irrational numbers are 'transcendental'), there is one more point I should like to make, arising out of the previous arguments. This is the importance of the purity of the methods used, to which the ancient Greeks paid so much attention. The question is not the general one of whether a square can be constructed whose area is equal to the area of a certain circle (at the end of the last century a mechanism was constructed which would turn out just such a square with perfect accuracy), but whether such a square can be constructed by means of ruler and compasses alone. In this sense the question has been definitely decided in the negative for all mathematicians. Only some poor fools do not believe this, and their imagination is tickled by the fantastic in the expression 'squaring the circle'.

Clear methodology, in other words the unambiguous statement of conditions of work, is the reason why mathematicians always understand each other so well, unlike workers in some of the other sciences. Mathematicians of all epochs and of all countries understand each other perfectly. Mathematicians are proverbial for their unintelligibility, although one can hardly imagine anyone who would clarify his statements with such meticulous regard for the other person as the mathematician. Of course even the subjects dealt with by Mathematics acquire a certain personal flavour peculiar to each mathematician in much the same way as is the case with other disciplines. For example the words 'point' or 'straight line' can mean something quite different to different people. One of our professors began his first lecture by asking one of the ladies: 'Madam, have you ever seen a point?' This was rather unexpected, but the answer came: 'No, I have not.' 'Have you ever drawn a point?' came the next question. 'I have,' came the reply, but the lady in question quickly changed her mind and said: 'I mean I have tried but have never succeeded.' (It is this answer that endeared our year to our professor for the rest of his life!) Those deposits of graphite or of chalk that we draw, and which under a microscope are veritable mountains, are, of course, not points. We all have some sort of an idea of a point, and it is this idea that we try to realize when we try to draw one. Our imaginings about straight lines can be even more personal. A straight line is not at all a simple line; little children and primitive savages never draw lines that are straight; what they draw spontaneously is a curve. In order to draw a straight line, it is necessary to possess self-discipline of a high order. For these reasons, if a mathematician has proved something about points and lines, he communicates his findings to his fellows as follows: 'I do not know what kind of pictures you have of geometrical figures. My idea is that through any two points whatever I can draw one straight line. Does this agree with your idea?' If the answer is in the affirmative, then he can proceed thus: 'I have proved something and during the proof I did not make use of any other property of points and straight lines apart from the ones about which we are already agreed. You can now think about *your* points and lines; you will still understand what I have to say.'

Mathematics does not pretend to enunciate absolute truths. Mathematical theorems are always put in the more humble form: 'If, . . . then . . .' '*If* we can use only ruler and compass, *then* the circle cannot be squared. *If* by points and lines we mean figures with such and such properties, *then* the following things are true of them.'

It is quite true that at school we were not used to these kinds of theorems, nor in the preceding chapters have theorems been expressed in these ways. Those who wish to convey knowledge do well not to convey it in a ready-made fashion, but rather in a kind of formative stage. Exact conditions are not readily formulated in the heat of their generation. Great constructive epochs are usually followed by critical epochs; mathematicians look back over the road travelled and try to get at the very kernel of the results themselves.

Euclid was one such great systematizer, and his works have survived and remained our models over the centuries. First he lists the fundamental ideas and the fundamental relationships between them (right up to the present day these have always been called axioms); the proofs that follow are only for those who imagine points, lines and planes in such a way that they accept the corresponding axioms as true. That is why the axioms are statements collected and chosen with great care so that everyone's perception should agree with them. For example one of the axioms states that, if we are given two points, we can draw one and only one straight line through them

His work is two thousand years old, and only one of the axioms has ever been the subject of debate. This is the famous

parallel axiom: you can draw only one line through a point not lying on a straight line, which does not meet that straight line, however far you produce the lines.

This one line which you can draw that does not meet the first line is what is called a line parallel to the first line.   We shall come back to this later.

I should first like to draw the reader's attention to another feature of the axiomatic method.   If the proofs for our theorems are such that we can all let our imagination run away with us as far as points, lines and planes are concerned, with the important proviso that our figures must satisfy the relationships embodied in the axioms, then it is really quite unimportant that these figures be points, lines or planes at all in any sense whatever. We might actually think of quite different objects, as long as these objects also satisfy the conditions embodied in the axioms, and our proof will lead to a true theorem about these objects. This again is a kind of 'I say one thing, then it turns into two things', which we came across when dealing with duality. The theorems in question would remain true even if there was a person with such a tortuous imagination that he thought of a straight line when we spoke about a point, and of a point when we spoke about a straight line.   (Perhaps the reader will recall the example quoted there: three points determine a triangle, as long as all the points are not on the same straight line: three straight lines determine a triangle as long as they are not all on, i.e. do not pass through, the same point.)

If for example somebody understands by a point any point inside a certain circle (excluding the points on the circumference), and by a straight line only those parts of straight lines that are inside this circle

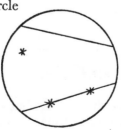

—even in this narrow world it is still true that through two points (i.e. through two points lying inside the circle) we can draw one and only one straight line (i.e. part of a straight line reaching as far as the circumference of the circle); even here all theorems will be true which can be deduced about points and straight lines by means of this axiom alone.

Now let us turn again to the parallel axiom.   I believe that anyone who thinks about it for a little while will agree that you can draw only one parallel to a given straight line through a given point, and will fail to see anything problematical about the question at all.   Most people's perception is indeed such that they accept the parallel axiom without question.

But I should like to tell the reader an experience I had while teaching a first form in a Grammar School.

Every pupil had a square in her hands, and the task was to say what could be noticed about the sides of the square. Soon the word 'parallel' was mentioned, since children come across this word in the ordinary way.   I asked them what they understood by the word parallel.   One little girl said that parallels have the same direction, another that parallel lines always remain at the same distance from each other, a third one that however far we produce them, they will never meet. 'This is all quite correct,' I told them.   'We could accept any one of these as the sign by which we recognize parallels, the other two will then follow from it.'   At this Anne, in the first row, stood up (she was the most profound thinker in the class), and said: 'It would not be a good idea to accept as our definition that they never meet.   I can imagine two straight lines which do not remain at the same distance from one another; they get nearer and nearer to each other, and yet they never meet.' She drew a figure on the blackboard as well, indicating what she meant:

I had to accept the fact that Anne's perception was indeed different.

The trouble is that things like that cannot be checked by experience.   If we bend our usual parallel down a little then

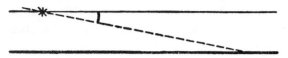

by producing the lines far enough we can still show that they will meet. But supposing we bend it down by much less, i.e. by $\frac{1}{10}$th, by $\frac{1}{100}$th, by $\frac{1}{1000}$th of the arc that we started with; we can continue this infinite sequence as long as we like and how do we know that we do not eventually reach an extremely small degree of bending which results in our bent line not cutting the lower line? The thing is we cannot go *right through* the infinite sequence.

We have already come across lines which get nearer and nearer a straight line without ever reaching it. Either branch of the hyperbola is like this.

It is not really surprising that there are people who can imagine straight lines getting nearer and nearer to each other in this way. Our imaginings are guided by our sensory experiences. It might be the case, for example, that for somebody separated for a long time from a loved one, the picture of getting nearer and nearer could develop in quite a definite form in his imagination, without the possibility of an actual meeting.

However all this may be, ever since Euclid there have been many people whose perception was like that of my pupil Anne. Probably they were not very sure about the products of their imagination, since the view of the majority was against them, but they still doubted whether the parallel axiom was as self-evident as the other fundamental truths. They would say: 'Why don't you prove it, using only such relationships as we can also accept. Then we shall accept it too.'

Through several centuries mathematicians tried to prove the parallel axiom with the aid of the other axioms, but without success.

The Hungarian John Bolyai was one of the first to take a stand by the kind of perception like that of my little pupil's. 'The reason that nobody has succeeded in proving the parallel axiom is that it is not true. I see the thing in this way: if I draw a line through a point outside a certain fixed line to

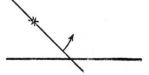

intersect this line then this intersection will move farther and farther away as I turn my line about the point; eventually there will not be any intersection at all:

but the turning line is still slightly bent towards the fixed one. Of course if I turn it round some more, it will be even less likely to intersect the other line until, of course, it begins to bend towards the fixed line on the other side:

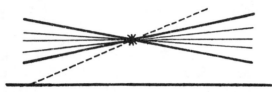

There are therefore two leading straight lines through the external point, and any straight line between these two will not meet the fixed straight line at all, and those that bend more towards it will all intersect it.   Let everyone join me who sees this in the way I do, and I shall construct our own Geometry.'

Bolyai took as his fundamental relationship the opposite of the parallel axiom, kept all the other Euclidean axioms, and investigated what kind of theorems could be deduced from these fundamental relationships concerning points, lines and planes.   In the Bolyai Geometry, constructed in this way, there are a lot of things which differ from Euclidean Geometry.   It is a matter of taste which of them we care to accept.

It does not detract in any way from Bolyai's merit (although it utterly broke the unfortunate man) that at the same time others also discovered the possibility of having different geometries.   This is quite a frequent occurrence: it seems as though certain problems somehow ripen through the passage of time, and there are people who are sensitive to this at different points of the world, and so simultaneous but independent discoveries are made.

There is still something not quite in order here.   May it not still be possible to prove the parallel axiom, and if so the whole of the Bolyai Geometry would be based on a false premiss, and eventually a whole host of contradictions might be deduced from it?

Fortunately we have a comforting answer to this awkward problem. From the point of view of reliability Euclid's and Bolyai's Geometries are as good as each other. If the Bolyai Geometry were to lead to contradictions, then Euclidean Geometry would also contain contradictions.

We can see this because it is possible to build a model of the Bolyai Geometry entirely within Euclidean Geometry. We have considered a world with a narrow horizon, whose points and lines all lie within a circle of Euclidean Geometry. There we showed that even these points and lines, understood in this narrower sense, satisfy one of the fundamental relationships in Euclid. It can also be shown that all the other fundamental relationships are satisfied (assuming we suitably transform the notion of congruence), with the single exception of the parallel axiom, instead of which we have Bolyai's fundamental relationship.

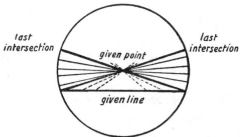

The 'leading' straight lines are those which lead from the given point to the extreme points of the given fixed straight line, i.e. to the circumference of the circle. The straight lines in between these (i.e. those parts within the circle) do not intersect the given fixed straight line, even according to Euclid. Bolyai's axiom cannot therefore contradict the other Euclidean axioms, since in this narrow world they can get along quite nicely side by side.

So we have now come across two geometries of equal standing; there is no reason why we should not now speak about geometri*es* in the plural. We could, as a matter of fact, go on playing this game quite independently of any perception: in the place of any one of the axioms which cannot be proved from the others, we could assume its contrary and investigate what kind of theorems could be deduced from this contrary hypothesis. Moreover, we could assume quite different axioms, since it does not seem worth sticking to axioms derived from perception; the Bolyai type of geometry has already shown us how unreliable a basis perception is. If everyone took notice of his own type of perception, quite contrary results could be obtained, as we have already seen.

It is in this way that a whole series of geometries were constructed one after the other. And this is not merely a game; modern Physics has recourse to just such abstract geometries to explain real events.

The perception of man is not unalterable. The development of science goes on shaping it all the time. When it was discovered that the Earth was not a flat disc and it had to be worked out how people on the other side of the world could be walking on their heads, man's perception immediately made great strides. If the results of modern Physics become more or less permanent and pass over into general knowledge, then in time people with Euclidean perception will perhaps cease to be in the majority, and one of the geometries which today appears as an abstract game may become the geometry of reality.

Postscript about the fourth dimension

I should like to come back once more to the idea of a 'model'. We were able to construct a model for the Bolyai type of geometry within Euclidean Geometry by circumscribing a part of a Euclidean plane by means of a circle. To every theorem in Bolyai Geometry corresponds a theorem that can be proved inside this circle. We already came across this kind of intertwining of two branches of a science when we found a model for Geometry in the form of Algebra. To points correspond pairs of numbers, to lines correspond equations with two unknowns, and we circumscribed that part of Algebra within which every geometrical figure represented an algebraical expression, and every geometrical theorem an algebraical theorem. In this way we are able to prove geometrical truths by algebraical methods and, conversely, we are able to make use of geometrical results in the examination of the properties of functions represented by curves.

All this was in a plane, but there is no reason why it should not be carried over into three dimensions, lock, stock and barrel. In three-dimensional space a point is determined by three numbers (if the bird's nest had been at the top of a tree, in order to determine its position exactly it would have been necessary to know how tall the tree was, i.e. what size ladder would be necessary to take along in order to reach it). To figures in space will thus correspond equations with three unknowns. We could denote the three unknowns by X, Y and Z. If we are dealing with an equation of the form

$$Z = 3X + 2Y$$

it can be seen immediately that the value of Z depends on the choice of X and of Y. Such functions are called functions of two variables. (We often come across such functions in everyday life; for example, the amount of a life-insurance premium depends on the length of time the policy is in force as well as on the capital sum insured.) Whatever we may prove about figures in three-dimensional space will be expressible in terms of functions of two variables.

Of course there is no need to start everything from the beginning just because we are dealing with three-dimensional space. The majority of theorems of plane geometry may readily be generalized to such a space. For example, in a plane the way to find the distance of a point from the 0 point is as follows:

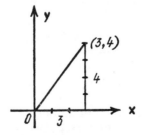

The distance required is the hypotenuse of the right-angled triangle the other two sides of which are the co-ordinates of the point in question. By Pythagoras' theorem the square of this distance is equal to the sum of the squares of the other two sides, and the distance itself will be

$$\sqrt{3^2 + 4^2}$$

It can be proved that the point in space characterized by the three co-ordinates (3, 4, 5) is at a distance

$$\sqrt{3^2 + 4^2 + 5^2} \text{ from the 0 point}$$

Quite often generalizations from plane to three-dimensional space are as easy as that. The effect of this is that a whole lot of theorems concerning functions of one variable may very simply be generalized to functions of two variables.

It is quite possible, of course, that we may come across functions of 3, 4, . . . or any number of variables. It seems a pity that whereas we can pass from a two-dimensional plane to a three-dimensional space, we cannot pass on beyond our three-dimensional space. There just is no four-dimensional space to which to pass. On the other hand, the algebraical model allows one to act as though there were one. Let us, for example, call a set of four numbers like (3, 4, 5, 6) a point and let us call the number

$$\sqrt{3^2 + 4^2 + 5^2 + 6^2}$$

its distance from the 0 point. We may work with these numbers in the same way as we worked with the numbers corresponding to actual points, and we may hit upon theorems that can be deduced about functions of three variables. We can, of course, check that theorems obtained in such a fictitious manner are in fact true, and so it turns out that it was worth pretending that there was a fourth dimension, even though in fact there is not one.

In the same way we can introduce abstract spaces of 5, 6, . . . or even an infinite number of dimensions. Our starting point is always our well-known three-dimensional space and our aim always utility in the investigation of the properties of functions.

These are no longer unfamiliar concepts. The multi-dimensional points are just 'ideal elements', they come to our aid from an imaginary world, and, if we wish them to, they can disappear again, leaving behind them solid results which remain true without the intervention of the imaginary objects.

19. The building rocks

ONE of the activities of the great critical epochs is the extrication of the kernels of results already obtained, the clarification of the conditions of theorems, in a word axiomatization. This also circumscribes each branch of Mathematics; we regard those parts of Mathematics as unified wholes which can be deduced from certain sets of axioms.

As we look back over the road travelled, we notice that certain ideas appear here and there, in other words there are ideas that do not get circumscribed even after an attempt at systematization. They turn up in all sorts of different branches of Mathematics. So we have come across another type of activity in which we could engage; we could separate and make into special objects of investigation those elements which turn up in widely separated places.

For example we might remember that in the case of rational numbers we could always carry out multiplications and divisions (apart from division by zero), and we always obtained a rational number as a result. In this sense rational numbers, if we leave zero out, form a kind of closed group in relation to multiplication and division. Whole numbers do not behave in such an exclusive fashion, for division definitely leads us out of the set of whole numbers.

Whole numbers and rational numbers are similar in the sense that in relation to additions and subtractions they both form closed groups; of course we must think of positive and negative whole numbers, but these operations do not in fact lead out of the sets of these numbers; it is not even necessary to exclude zero.

There is naturally no need to have so many numbers in order to form a group which is closed in relation to certain operations. Supposing we consider only the two numbers

$$+ 1, - 1$$

We can multiply and divide these numbers as long as we like, the result can never be other than either $+1$ or -1.

This sphere of ideas is not even restricted to operations with

225

numbers.   Let me remind the reader of vectors, for even these form a closed group in relation to their queer kind of addition. The combined effect of two vectors is again some other vector

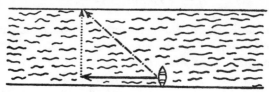

Such an operation can be called an addition only in a figurative sense.   What we are really talking about is the combination of movements and of forces.

One could go on for a long time giving similar examples.

The investigation of the 'group' idea, which seems to occur in many different places, in an independent sort of way, i.e. the theory of groups, has proved most fertile.   It is the essence of modern Algebra, and is made use of by modern Physics. The various geometries can be regarded as theories corresponding to different groups.

Groups themselves are 'sets' with certain particular properties.   This idea of 'set' is again one that we come across all the time in many different branches of Mathematics.   Whenever we speak about Mathematics, it is almost inevitable that we speak about sets of points, sets of numbers or sets of functions of a certain type.

It was Cantor who made this idea the object of his investigations.   The 'theory of sets' was really largely his creation.

Let us go back a little.   We spoke about the set of rational numbers and about the set of points corresponding to it on a line, and we decided that every single point of this set is a 'point of condensation'.   This is a most important idea in the theory of sets of points.   We call a point a point of condensation of a set if, even in the closest vicinity of the point, there are always other points of the set.

We have also seen some of the methods employed in the theory of sets.   Let us refresh our memories by another example.   There are an infinite number of natural numbers

$$1, 2, 3, 4, 5, \ldots$$

and yet they are not condensed anywhere, they go marching on for ever by unit steps. But let us stuff a whole infinite set into a finite interval; for example

$$1, \tfrac{1}{2}, \tfrac{1}{3}, \tfrac{1}{4}, \tfrac{1}{5}, \cdots$$

are all within the interval between 0 and 1

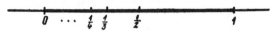

so there must be a condensation point somewhere within this interval.

This can be proved quite generally as follows. Let us suppose that all points of an infinite set lie between the points 0 and 1

and it does not matter exactly where they are. Let us halve our interval. At least in one of the halves there must be an infinite number of points of the set, because if each half had only a finite number of points in it, say 1 million in one half and 10 million in the other half, this would be 11 million altogether, quite a big number, but still finite. In our previous example the infinite number of points lie in the left-hand half of the interval.

Now, instead of the original interval, let us consider the half that has an infinite number of points, or either half if they both have an infinite number of points. Suppose that the new interval is:

We can repeat the previous argument exactly about this interval; we can proceed to one of its halves, i.e. that half in which there are an infinite number of points of the set. This halving can be continued indefinitely. In this way we obtain intervals encased in each other and getting smaller and smaller. In our example we should obtain

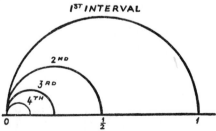

It can easily be seen that the lengths of these intervals converge to zero. Here again we are dealing with the amusing parcels consisting of wrapping after wrapping, and at the common centre of them all is a screwed-up piece of paper. We can see in our case, too, that there will be one single point common to all our intervals. This point must be a condensation point of our set, since within an arbitrarily close vicinity of it will be found some of our intervals which have shrunk even smaller, and each one of these contains not only one but an infinite number of points of the set.

Now we have reached such dizzy heights of knowledge that we can even answer the question of how a mathematician catches a lion. The method employed by experimental physicists for catching lions is well known. It can be understood by any beginner and applied. The experimental physicist pours the whole of the Sahara on to a sieve, the part that goes through the sieve is the Sahara, the part that is left is the lion. The mathematician, on the other hand, proceeds quite methodically as follows:

It is necessary to distinguish two cases:

Case (1). The lion is at rest.

We must prepare a cage, open underneath, which is large enough to hold the lion. Now divide the Sahara into two equal parts. The lion will be in at least one of the halves (since if it is on the boundary, it will be in both parts). Now let us consider such a half-Sahara. Let us divide this into two halves. Our lion will be resting in at least one of the halves. And so we continue these halvings and obtain areas enclosed in each other, getting smaller and smaller. Sooner or later one of these areas will be smaller than the base of our cage, and our lion must be in that area too. Let us now place the cage on top of the lion. We have caught the lion.

Case (2). The lion is moving.
This method is then not applicable.
This is the point.

So much for point sets.

We have already come across proofs in the theory of sets which are not valid for point sets only. For example, the method of pairing by means of which we proved that the set of natural numbers and the set of rational numbers are equally numerous (though the irrational numbers are more numerous than the rational numbers), may be used for any kinds of sets. If I remember rightly, it was from the sets of dancing boys and girls that we finally passed on to these less frivolous sets. Anything we can say about how numerous sets are could be true equally for dancing couples, for real numbers, or, come to that, for the set of all the sentences that can be written down in the English language. Cantor dealt with sets in just such general terms. He proved a whole lot of remarkable theorems about the 'numbers' of infinite sets, i.e. about the extension of the finite-number concept to infinity. He showed for example that there were not just two kinds of numbers, that of the natural numbers and that of the real numbers; as a matter of fact there is no set, of whatever number, which cannot be transcended by another set of higher number. The poet Babits called these numbers, towering ever higher and higher, 'the towering battlements of infinity'. Cantor introduced certain operations for these numbers, namely additions and multiplications, somehow imitating the operations on our tiny numbers. Now this is really what we might call playing on a grand scale, playing with infinity. It appeared that the human spirit had reached the greatest heights of which it was capable.

And this is the point when the whole building started to rock.

At the end of the last century certain contradictions turned up inside Mathematics, inside this science which had always been considered almost boringly safe. And it was just where it had reached the greatest heights, in the theory of sets, that the Achilles heel of Mathematics came to light.

Out of all the contradictions let us go over the most serious one, Russell's antinomy. Let us first consider it in its more jocular form, in which it is generally known.

We can define the army barber as follows: he is that member of the army who is obliged to shave all those in his company who do not shave themselves, for, in order to save time, he is not allowed to shave those that do shave themselves. The question is whether this soldier does or does not shave himself.

If he does, then he is one of those that shave themselves and he is not allowed to shave such people.

If he does not, then he belongs to those that do not shave themselves, and he is obliged to shave all such people.

What is he going to do?

Of course in a joke like this the way the problem is expressed is inaccurate.   Let us now pass on to the more serious example.

A set is not usually an element of itself.   For example, the elements of the set of natural numbers are numbers, not sets, so the set itself, being a set, cannot belong among its own elements.

It could, of course, happen that the elements of a set contain sets.   For example, let us imagine all possible sets of numbers, and consider the union of all these sets as a single set.   One element of this set is, for example, the set of natural numbers, another all the numbers less than 10 and so on.   Every one of its elements is a set.   But it still is not contained among its own elements, since its elements are all sets of numbers, and the whole set itself is a set of sets.

Now if we try to unite all imaginable sets into one single set, then we have an example of a set which is itself one of its own elements.   Obviously so, since the whole united set is itself a set, and every set must occur somewhere among its elements.

Those who feel that it is going to be rather a bore to think all this out need not bother to do so.   There will be no further use made of it in what follows.   It is enough to take the point of view that in ordinary sets such oddities do not occur.   So let us call every set 'ordinary' if it does not itself figure among its own elements; there is no need to bother at all with the question of whether there are any other sets in reality.   Let us then imagine all 'ordinary' sets heaped together into one big set.

The problem is whether the set so obtained is 'ordinary' or not.

If it is 'ordinary', then it must figure among the elements of the big set, with all other 'ordinary' sets.   Yes, but this would make our big set not 'ordinary'!

If it is not 'ordinary', then it cannot figure among the elements of our big set, since these elements are all 'ordinary'. But this situation is just what we have called being 'ordinary'!

So, if it is 'ordinary', then it is not 'ordinary', and if it is not 'ordinary', then it is 'ordinary'.   Anyhow we have arrived at a contradiction.

And this cannot be helped.

It is equally useless to say that the theory of sets was in too much of a hurry to rise to great heights; let us drop the whole thing and go back to the humbler and safer branches of Mathematics. We know by now how the theory of sets was created: its ideas are to be found in every branch of Mathematics. If there were trouble in the theory of sets, then there might be trouble anywhere.

The repercussions of this shock are with us to this day.

Mathematicians take the sort of attitude in the face of this situation that people would usually take in the case of any prolonged danger. Most of them do not even want to think about it, everyone carries on with his own job, and if anyone happens to mention the danger it is usually waved aside with a little nervous protest.

There are of course a few who are trying to save the situation.

Of course, at first the error was sought in Russell's antinomy itself. Russell himself thought that the very definition of the set occurring in the antinomy was wrong. The definition is a 'vicious circle', since the set to be defined is included in the definition. We could unite all 'ordinary' sets into one set only if we could already decide beforehand of the set so generated whether it is ordinary or not, and so whether it can be accepted as an element at all.

Unfortunately we find such 'vicious circles' all over the place in every branch of Mathematics. In the case of natural numbers it is quite usual to give definitions like this one: 'Let us consider the smallest number which has such and such properties.' This number, too, figures in its own definition. We can choose only the smallest out of *all* the numbers with the said properties, and the smallest one is bound to be among all of them.

The most radical rescue attempt is that of the intuitionists (the term 'intuitionism' is not a very fortunate one here, but we shall not trouble about its exact meaning). The line they take has a history longer than that of the antinomies, but the antinomies have given its adherents a new lease of life. The new intuitionism is coupled with Brouwer's name. He rejects the whole of Mathematics as conceived up till now, and makes an attempt at building it anew on more secure foundations.

He accepts only that which can in some sense be constructed, since once we have constructed something, it is undeniably there, no antinomy can ever make it non-existent.   He rejects the 'proofs of pure existence', for example the old proof of the fundamental theorem of Algebra, since it gives no method for the construction of the roots of the equation.   He will have no truck with 'actual infinity', since we can actually construct only a finite number of elements of any set, even if this can be continued indefinitely.   An infinite set according to him is only 'potentially infinite', it is always in the state of being generated, and can never be considered as finished or closed.

Thus only the ruins of classical Mathematics remain in this way, and what remains becomes terribly complicated by reason of the necessity for carrying out the constructions in every case.

Only Hilbert's rescue attempt can be regarded as realistic. The significance of this attempt has grown beyond its original object of coping with the above-mentioned dangers.   A new and fertile branch of Mathematics has grown up out of it. We shall discuss these developments in what follows.

20. *Form becomes independent*

THE reader must not imagine that the theory of sets is still burdened with the weight of the antinomies. When the time came (it is the contradictions that made things so terribly urgent) to put the original naïve theory of sets in proper order, to establish a system of axioms for it, mathematicians took great care to restrict the idea of set sufficiently by means of the fundamental conditions stipulated. They succeeded in keeping all that is valuable in the theory of sets, and the troublesome sets were left outside. But this seems a very artificial kind of order; as Poincaré has said, we have built a fence round the flock to save the sheep from the wolves, but how do we know that there are no wolves hiding in certain places inside the fence? There is no security against further contradictions turning up.

One of the greatest mathematicians of our time, Hilbert, set himself the task for the last twenty years of his life of looking into every nook and cranny inside the fence of the axiom system. He acknowledged that we may justifiably be worried about definitions involving vicious circles, about 'pure existence proofs' and about 'actual infinity'. There might be some danger lurking in any one of these. But why is it that we feel compelled to work with such dangerous 'transfinite' concepts which seem to be beyond our finite minds? There is a very good reason; and, except for extremely compelling reasons, we shall never want to do without these concepts for it is they that enable us to build comprehensive theories, since they make possible the discovery of connexions between far-distant territories. This is shown very well by the Mathematics of the intuitionists which falls into so many little separate pieces. Therefore we are unwilling to give up the dangerous concepts which weld the whole of Mathematics into one single powerful edifice.

The transfinite tools play the same kind of part in logic as the line at infinity or 'i' play within Mathematics itself. We may regard these as the 'ideal elements' of logic. We need to treat them in the same way as the mathematical ideal elements:

233

introduce them if they prove useful (and how useful they have proved to be!), but examine them very carefully whether they might be in contradiction with our established procedures. It follows that the task that remains is the examination of the freedom from contradictions of the transfinite processes.

So Hilbert's programme is the mathematical examination of logic itself as applied to Mathematics, i.e. of deductions, demonstrations, etc. A precondition of doing this is to cleanse these ideas of any vagueness which might attach to them on account of inexact linguistic expressions, and to extract from them their unambiguous pure form.

It was possible to begin to examine numbers in any exact way only when, after ceasing to speak about 5 fingers or 5 apples or 5 sentences, we considered the pure form which all these had in common. This is what we called their number and denoted by the sign 5. If we wish to examine statements, we must then disregard their content. For example what is of interest to us in such statements as '2 × 2 = 4', 'we can draw just one straight line through two points', 'snow is white', is what is common to them all. This is of course the fact that they are true. We can introduce a new sign for this, for example ↑. The common logical value of the statements: '2 × 2 = 5', 'two straight lines meet in two points', 'snow is black', is that they are all false. The sign for this could be ↓ (like an upturned thumb, which meant life, and a thumb pointing downwards, which meant death in the ancient Roman circuses).

In Mathematics we are interested only in statements which assume one or the other of these logical values (in other words those that are true or else false).

Here we are then about to construct a kind of arithmetic which is much simpler than the arithmetic of natural numbers. There is an infinite number of natural numbers, but here there are altogether only two values. It will be very easy to write down the tables for the various operations.

We are really going to deal with calculation; there will even be logical operations. These will be the connexions between different statements, connexions that we come across all the time in Mathematics.

Any mathematician can easily discover what these connexions

are, in a very simple way, as long as he does not speak every language.   All he needs to do is get hold of a book on Mathematics written in a language that he does not know and take note of the words that he is forced to look up in the dictionary while he is reading it.   He will be surprised to find that when he has learnt the expressions:

'not', 'and', 'or', 'if, . . . then', 'then and only then', 'all', 'there exists . . .', 'the one which . . .'

after a little while he will no longer be aware of reading a foreign language, for he will understand everything perfectly.   The formulae are, of course, in any case international, the text merely serves to bring out points of emphasis, and this is not absolutely indispensable.   The necessary logical connexions, on the other hand, are in the few expressions enumerated.

What is for example the table for the word 'not'?   It is extremely simple.   The negation of a true statement (for example, '2 × 2 is not equal to 4') is clearly false; the negation of a false statement is equally clearly true (for example, '2 × 2 is not equal to 5').   So the table for 'not' is as follows:

$$\text{not} \uparrow \; = \; \downarrow$$
$$\text{not} \downarrow \; = \; \uparrow$$

It is usual to abbreviate the word 'not' by the following sign:

$$\sim$$

For example $\qquad\qquad \sim (2 \times 2 = 5)$

is the negation of $2 \times 2 = 5$.   With this notation the table for 'not' is the following:

$$\sim \uparrow \; = \; \downarrow$$
$$\sim \downarrow \; = \; \uparrow$$

We can equally easily write down the table corresponding to the logical operation 'and'.   If two statements are true and we connect them with an 'and', we again get a true statement. For example '2 × 2 = 4, and only one straight line can be drawn through two points' is certainly true.   Therefore

$$\uparrow \text{ and } \uparrow \; = \; \uparrow$$

But, on the other hand, if only one of the connected statements is false, it ruins everything.   The statement '2 × 2 = 4 and

2 × 3 = 7' is a false statement, in spite of the fact that it has a true portion. Of course, connecting two false statements by means of 'and' is going to falsify all the more surely. So the table can be continued as follows:

$$\uparrow \text{ and } \downarrow \;=\; \downarrow$$
$$\downarrow \text{ and } \uparrow \;=\; \downarrow$$
$$\downarrow \text{ and } \downarrow \;=\; \downarrow$$

We have thus exhausted all the possibilities. It is a nice finite multiplication table, much simpler than any algebraical multiplication table.

What about finding the table for 'or'? We must make it clear at first what kind of 'or' we are thinking about. Linguistic expressions are in this connexion quite ambiguous.

'Either we are all mad and will perish to the last man,
  Or our Faith will be vindicated in the World'.

One of these will happen without a doubt, but both cannot happen. They are mutually exclusive.

'If we divide the Sahara into two halves, our lion will lie either in one half or in the other.' It will certainly lie in one of the halves, but it might lie in both if it happens to be having a stretch right on the boundary.

'A person is either eating or speaking', these two are mutually exclusive, but neither needs to be necessarily the case. It is possible to do something else with our mouth, for example, we need not open it at all.

In Mathematics the word 'or' is mostly used in the second sense. In other words, statements connected by means of the word 'or' can be regarded as true if at least one of the statements is true. We include the case when they are both true, but exclude the case when neither is true. Accordingly the table for the word 'or' is

$$\uparrow \text{ or } \uparrow \;=\; \uparrow$$
$$\uparrow \text{ or } \downarrow \;=\; \uparrow$$
$$\downarrow \text{ or } \uparrow \;=\; \uparrow$$
$$\downarrow \text{ or } \downarrow \;=\; \downarrow$$

Once we have obtained our table, we may regard the table as the definition of the logical operation 'or'. We have purged the word of its linguistic vagueness, the connexion will henceforth never be ambiguous.   The other two 'or's, i.e. those taken in the other two senses, may then be defined equally precisely, making use of our 'or'.

It is obvious that there must be some rules of manipulation here, too.   For example, in both the operations 'and' and 'or', the order of the two statements is interchangeable, just as the factors in a product.

I do not want to exhaust this subject entirely, although it would not take long to go through all the possibilities, for, with only two logical values, there cannot be very many.

Instead, I should prefer to show the reader how it is possible to play at arithmetic here.   For example we know that we can multiply the powers of the same base by adding the exponents, and in this way we reduce a multiplication to an addition. Perhaps there are such relationships amongst the logical operations as well?

Let us take an example from the favourite topic of detective fiction.   Let us try to work out who the murderer is from the following set of facts:

In a murder case there are two suspects, Peter and Paul. Four witnesses are examined.   The first witness testifies as follows:

'All I know is that Peter is innocent.'

The second one as follows:

'All I know is that Paul must be innocent.'

The third one as follows:

'I know that out of the last two depositions *at least* one is true.'

The fourth one finally says:

'I can state categorically that the third witness uttered a falsehood.'

Upon examination of the facts of the case it transpires that the fourth witness was right.   Who is the murderer?

Let us analyse it, going back step by step.   The fourth deposition has been proved correct, so that the third witness did in fact testify falsely.   So it is not true that out of the first two depositions *at least* one is true.   Neither can be true.   It cannot be true that Peter is innocent, or that Paul is innocent. They are both murderers.   They must have been accomplices.

Now let us try to get at the logical kernel of the argument. It is quite useless to know what the depositions are, we must regard their logical values as unknown, since we do not know whether they are true or false.   Let us call the logical values of the first two depositions X and Y respectively.   The third witness testified that out of these two at least one is true, and (since our 'or' operation expresses just this sort of 'at least'), symbolically,

$$X \text{ or } Y$$

is a true statement.   The fourth witness denies this, the sign for denials is $\sim$, so according to him the truth is

$$\sim (X \text{ or } Y)$$

When we thought the matter over, we concluded that this was the same as saying that the truth must be the contrary of both the first and of the second depositions, so that the truth really is

$$\sim X \text{ and } \sim Y$$

The content of the argument is that whether X and Y are true or false, the statement

$$\underline{\sim (X \text{ or } Y)}$$

is completely equivalent to the statement

$$\underline{\sim X \text{ and } \sim Y}$$

and in this way we can pass from an 'or' connexion to an 'and' connexion and vice versa.

Of course the road to such relationships is not as a rule through any kind of joke.   Their truth can be checked quite mechanically.   We can write for X and for Y the values $\uparrow$ and $\downarrow$ respectively, and see whether the above two statements always yield the same value.   There are altogether four possibilities to try out:

(1) Both X and Y have the value $\uparrow$
(2) X has value $\uparrow$, but Y has value $\downarrow$
(3) X has value $\downarrow$, but Y has value $\uparrow$
(4) Both X and Y have the value $\downarrow$

Let us try the first case.   What will be the value of the statement

$$\sim (X \text{ or } Y)$$

if both X and Y have the value $\uparrow$?   According to the table for 'or' (do not trouble to think about it, just look it up)

$$\uparrow \text{ or } \uparrow = \uparrow$$

so that in this case

$$X \text{ or } Y = \uparrow$$

so we are faced with a statement

$$\sim \uparrow$$

but according to the table for 'not', its value is

$$\underline{\underline{\downarrow}}$$

Now what will be the value of the statement

$$\sim X \text{ and } \sim Y$$

if both X and Y have the value $\uparrow$?   In this case

$$\sim X = \sim \uparrow \, = \downarrow$$

and $$\sim Y = \sim \uparrow \, = \downarrow$$

so we are dealing with the statement

$$\downarrow \text{ and } \downarrow$$

and, according to the table for 'and', this is likewise

$$\underline{\underline{\downarrow}}$$

In the same way it can be shown that, in the other three cases, the two statements considered have always the same value.

We can play at Algebra too.   We can think of a statement, carry out all sorts of logical operations with it, and finally say whether we have obtained a true or a false statement.   We ask someone to find out whether we started off with a true or with a false statement.   The following type of game has particular importance here:

'Think of a statement, connect this by means of 'or' to its own negation.   There is no need to give anything away.   I shall still know that you must have obtained a true statement.' We can write it down in the following way: the statement thought of is X, its negation is $\sim X$, their connexion by means of 'or' is

$$X \text{ or } \sim X$$

and our contention is that the value of this statement is bound to be $\uparrow$, whether X was $\uparrow$ or $\downarrow$.   Let us try it out.

If the value of X is $\uparrow$, then by the table for 'not'

$$\sim X = \sim \uparrow = \downarrow$$

so we are dealing with the statement

$$\uparrow \text{ or } \downarrow$$

The reader can verify from the table for 'or' that the value of this statement is in fact $\uparrow$.

If, on the other hand, the value of X is $\downarrow$, then $\sim X = \sim \downarrow = \uparrow$ by the table for 'not', so we are dealing with the statement

$$\downarrow \text{ or } \uparrow$$

and, according to the table for 'or', the value of this is again $\uparrow$.

There are therefore connexions between statements which are always true, quite independently of the statements occurring in them, i.e. independently not only of the contents but even of the logical values of the statements. They are true entirely by virtue of their logical structure; they are called logical identities. These are statements that play a crucial role in Mathematics.

We can go on playing this game by pretending that the whole statement is not unknown, we just do not give away the subject. For example: 'I have thought of a number, and I state that it is an even number. Now I shall perform some operations on this statement.' We can write down the statement as follows:

$$\text{'}X \text{ is even'}$$

Whether it is true or false naturally depends on what X is. For example if $X = 4$, then it is true; if $X = 7$, then it is false. So we are dealing with a statement whose value is a function of X. We have therefore reached the theory of logical functions.

There is no reason why we should not consider logical functions of several variables. 'I have thought of three points and I state that they all lie on the same straight line.' We can write this statement in this way:

$$\text{'}X, Y, Z \text{ all lie on a straight line'}$$

and its logical value depends on the choice of the points X, Y, Z. If we choose three points in this way

$$\underset{x}{*} \qquad \underset{y}{*} \qquad \underset{z}{*}$$

then it is true; if we choose them like this

then it is false. We must take care here that the unknowns are not chosen quite arbitrarily. In the first example we had to choose among the natural numbers, in the second example X, Y, and Z had to come from the points of a plane or of a three-dimensional space. But we are already familiar with this from the sphere of mathematical functions. In those cases it was also necessary to specify from what set the unknowns could be selected. Here this set is called the 'universe of discourse' of the function in question.

Now we must introduce the dangerous operations. We apply them to these logical functions. Such is, for example, the little word 'all'. Let us apply this operation to our first logical function:

'For all X I state that X is even'

(naturally X is understood to have been chosen from among the natural numbers). We certainly obtain a statement, although a false one, since we can instantly think of an example to the contrary, for example 5 is not even. Therefore

'X is even for all X' $= \downarrow$

On the other hand, if we apply the words 'There exists' to our function, we obtain a true statement:

'There exists an X for which X is even'

and so

'There exists an X for which X is even' $= \uparrow$

We can see that the words 'all' or 'there exists' signify logical operations which we can apply to logical functions and obtain statements with definite values. In our example the statement beginning with 'all' had quite definitely the value $\downarrow$, and quite independently of X, and the statement beginning with 'there exists' had the definite value $\uparrow$.

These new operations have brought with them the transfinite elements. 'Something is true for *all* elements of the universe of discourse'; if the universe of discourse is infinite, as is the case with natural numbers or with the set of points in a plane, then we talk about the infinite as though it were something finished and closed in our hands.   'There is an X in an infinite universe of discourse', as though we could look right through this infinite universe and find in it the X we are looking for.   The previous statements are statements about the 'actually infinite' and 'pure-existence statements' respectively. 'There exists' statements state something about an element without being able to produce that element in fact.   This is how the 'ideal elements' come into Logic, which can acquire civic rights only after proofs of non-contradiction.

The statements of the theory of logical functions can be formulated in just as exact a way as the identity

$$X \text{ or } \sim X$$

In order to eliminate any ambiguity which might creep into these purely logical statements, owing to language being inexact, it is better to introduce signs instead of the ambiguous words in general use, as we have already done in the case of the word 'not'.   This is how the internationally comprehensible books of symbolic logic are born, in which on page after page not a single word occurs; they are mainly filled with symbols. The professional will read the meaning of the symbols in the same way as a musician will hear the tune when he reads the music.

Leibniz initiated the construction of a pure and unambiguous logical sign-language.   Many people have since developed it further, until finally Hilbert, with his colleague Bernays, fashioned it into today's fine, flexible instrument, which gives the deductive methods of Mathematics such an exact form that they can become objects of mathematical investigation themselves.

21. *Awaiting judgement by metamathematics*

IT is now time to consider a well-circumscribed branch of Mathematics and to establish whether there can be any contradictions in it.

We know already what the methods are for such circumscribing: we must somehow get at the fundamental conditions satisfied by the relevant theorems, i.e. the axioms, and then we can say that this branch of knowledge consists of what can be deduced from these axioms.

We can write down the axioms in the language of symbolic logic and so they will consist of a certain succession of mathematical and logical symbols, without any possibly ambiguous words intruding themselves.

There is one more matter we must examine carefully. What do we mean when we say that something can be deduced from the axioms? In other words we must clarify quite precisely the steps of a deduction.

When we deduce the correctness of a statement from the correctness of another statement, and write this in the language of our symbols, what happens is that we pass from one succession of symbols to another. Let us go back for a moment to the solution of equations, which also consisted of some such steps. For example, it was useful to pass from the succession of symbols

$$\frac{5X}{2} + 3 = 18$$

to another

$$\frac{5X}{2} = 15$$

We thought about this quite carefully before doing so, saying that if after adding 3 a number became 18, the number must be 15. But afterwards we noticed that *formally* the only difference between the successions of signs was that in the first succession there was a term 3 to be added, and there was no trace of this in the second succession, whereas the number on the right-hand

243

side was 3 less in the second than in the first succession. We deduced the purely *formal* rule that it is permissible to take a term to be added from one side of an equation to the other as a term to be subtracted, and afterwards we made use of this rule without giving any more real thought to the matter. The deduction which was thought out by the consideration of its content became in this way a mechanical 'rule of the game'. 'Certain symbols can be moved here and there with certain alterations.' This is like the rules in the game of chess: e.g., the king can move one space in any direction.

This is what can be done in a quite general way when we make further deductions from our axioms. We observe what *formal* alterations in the succession of symbols correspond to the deductions, and then we employ this formal alteration without any considerations of content.

After all this we can even forget with what the particular branch of knowledge is concerned, and we can say something like this: We have a few meaningless successions of symbols (we shall call these axioms) and a few rules for the game, which tell us to what successions of symbols we can pass from a given succession (these will be called rules of procedure or of deduction). This system of theorems and of proofs has indeed become such pliable and obedient material in the hands of mathematicians as have the numbers themselves. The well-established mathematical procedures can now be applied to this material.

These procedures, on the other hand, must on no account be applied mechanically as rules of the game. Every single step must be carefully weighed up: is this really an undoubtedly permitted form of deduction, have any dangerous elements got in by the back door? The aim must not be lost sight of for one moment: we want to justify the use of transfinite elements in a certain branch of knowledge and there would be no point at all in such a justification if the dangerous elements crept into the justification itself. Tools must be kept absolutely pure, so pure that the most rabid intuitionist cannot take exception to them.

This is where Mathematics is split in half. In one half there are completely formal systems, instead of deductions there are formal rules of procedure; the other half is a kind of super-

Mathematics, known as metamathematics, which carefully weighs up the content of every single step and uses only deductions free from danger; it somehow examines the formal systems from above, and its principal aim is the demonstration of the freedom from contradiction of different branches of knowledge.

But if we want to find out whether we can arrive at contradictions by using our rules of procedure, is it not then necessary to examine also the contents of the statements of the system? We would naturally imagine that it was the contents of the statements and not their form which might lead to contradictions.

We are helped over this difficulty by the fact that it is sufficient to take into account one single contradiction; for example, (if the natural numbers belong to the system) this one:

$$1 = 2$$

This simple succession of symbols can be remembered as a succession of formal signs. We notice that a succession of a 1, the symbol $=$, and a 2, means a contradiction. There is no need for anything else. We have come across jocular ways of proving that $1 = 2$, and I informed the reader at that point that once we smuggled in a single contradictory statement, then everything becomes provable, in particular $1 = 2$. It is therefore enough to show that the single formula $1 = 2$ is not capable of being deduced within the system. In that case it will be quite certain that no contradictions can have crept into it in any other way.

The problem for metamathematics, formulated with perfect exactitude, is this: to show that, starting from the initial successions of symbols called axioms of the system, we can never reach the succession $1 = 2$ by using the given rules of procedure.

Hilbert himself showed in certain simple cases how to prove such freedom from contradiction, and a number of his pupils then generalized the procedure to wider systems. The first in the field, even before Hilbert, was actually Gyula König, the man who transplanted into Hungary almost all the branches of modern Mathematics.

We were now quite ready to examine a whole extensive

branch of knowledge; we had all the tools ready. The first branch to examine would seem quite naturally that of the natural numbers. Everything seemed to indicate that by a little concentration of forces Hilbert's ideas could be extended to the whole theory of numbers, including all the inherently dangerous ideas.

Then something else happened. Hilbert's 'theory of proof', this slowly and carefully growing new branch of science, was shaken by another storm.

A young Viennese mathematician called Gödel, making use of exactly the methods of the theory of proofs, proved that the freedom from contradiction of the theory of numbers cannot be proved by means of tools which are themselves expressible formally within the system. (The way he proved this will form part of the last chapter).

Let us make sure that the meaning of this is clear. Metamathematics does not make use of formal tools; if we are working in metamathematics it is necessary for us to know exactly what we are doing, we must make our deduction consciously, not mechanically. It does not follow, of course, that it would necessarily be impossible to turn these deductions into formal rules of procedure. This would naturally be quite possible for someone who wanted to play with these ideas independently of the aims which metamathematics had set itself. To do this it is not even necessary to be John Neumann, whose saying has become proverbial: other mathematicians prove things they know, Neumann proves what he wants to prove. (He is reputed to have said at a Congress in Bologna that the formalization of metamathematics was not interesting, but that he would do the whole thing himself for a box of chocolates). *If* we did formalize metamathematics, it would seem self-evident that its methods of deduction, built carefully to avoid all dangerous elements, should be capable of being formalized in a much narrower framework than the branch of knowledge under examination with its transfinite elements. But no, Gödel's result tells us that this freedom from contradiction can be proved only by means of methods which go beyond the methods of the system examined. Who is going to be satisfied with a justification of the dangerous elements by means of methods taken from a sphere which is wider than the system? It seemed that the theory of proof had failed, that we could shut up shop and go home.

Hilbert himself did not believe this for a moment. He was convinced that there was a way out. There must be some method of deduction which slips out of the framework of the examined systems, and yet builds on some concrete capacity of our finite brains in such a way that even the intuitionists would accept it.

The search for such methods of deduction was begun immediately, and the search was crowned with success. Gentzen found the required tool for metamathematics in the form of 'transfinite induction', and with the aid of this tool he did in fact manage to prove the freedom from contradiction of the whole of the theory of numbers. The flock of natural numbers may live and grow in peace; no wolves can possibly ever turn up amongst them.

'Transfinite induction' sounds very dangerous. In fact what it means is expressed by the following innocent argument.

If we start at any member of the sequence of natural numbers

$$1, 2, 3, 4, 5, \ldots$$

however far away, and take arbitrary steps backwards, it is quite certain that we can take only a finite number of steps. Even if we start at 1 million, and amble back in steps of a unit length, we shall reach 1 after a million steps.

Now let us rearrange the sequence of natural numbers, for example, by taking the odd numbers first and the even numbers afterwards:

$$1, 3, 5, 7, \ldots \quad 2, 4, 6, 8, \ldots$$

If we walk backwards in this arrangement, i.e. if we go on choosing numbers always nearer and nearer to the beginning, our choices will necessarily come to an end after a finite number of steps. In fact, if we start with an odd number, then the thing is just as obvious as in the original sequence of natural numbers. If we start with an even one, we can see in the same way that, going backwards, sooner or later we shall run out of even numbers, and after this we can choose only an odd number. The moment we leap over to the odd numbers, however big a number we choose, we are then moving in one single sequence, and this is just like the sequence of natural numbers in their original arrangement.

It is, of course, possible to rearrange the sequence of natural numbers in a much more complicated way. For example, we could enumerate the natural numbers by splitting them into groups as follows: numbers divisible by 3, the numbers that are 1 more than the numbers divisible by 3, the numbers that are 2 more than the numbers divisible by 3 (let us include 0 for the sake of tidiness):

$$0, 3, 6, 9, \ldots, 1, 4, 7, 10, \ldots, 2, 5, 8, 11, \ldots$$

If we start off with any of the numbers in the third group, we must leap over to the second group after a finite number of steps, and then the situation is the same as in the case just considered.

It is possible to obtain an infinite number of groups by, for example, separating off the odd numbers, then those which are divisible only by the first power of 2, then those that are divisible by the second power $2^2 = 4$, then those that are divisible by the third power $2^3 = 8$ and so on:

$$1, 3, 5, 7, \ldots, 2, 6, 10, 14, \ldots, 4, 12, 20, 28, \ldots,$$
$$8, 24, 40\ 56, \ldots$$

We need not be afraid of having an infinite number of groups, for the moment we start off with a definite number, this must be in one of the groups which can be preceded only by a finite number of groups.

We have seen that in each case walking backwards really involves passing from a more complicated arrangement to a less complicated one. So it should also be clear that if we start from any arrangement of the sequence of numbers, however complicated, and pass over to less and less complicated arrangements, a finite number of steps will take us to a simple sequence without any complications.

What Gentzen makes use of in his proof is the fact that, starting from an arrangement considerably more complicated than any just mentioned, there are still only a finite number of steps to take us back to the beginning. This is a kind of statement quite easily conceivable to our finite minds, yet it slips outside the framework of the system considered.

How is it possible to use this tool in a proof of non-contradiction?

An argument to show non-contradiction usually goes something like this: suppose that someone has found a contradiction which can be deduced from the axioms of the system. He hands over a proof, which starts off with the axioms, proceeds by means of the admitted rules of procedure, and ends with $1 = 2$. We have to show that this proof is fallacious; in fact we must find the flaw in it.

If not a single dangerous element occurs in the proof, then it is clear that we can find the flaw in it. If our starting point is correct, then generally accepted methods of proof could lead anyone to the result $1 = 2$ only if he has made a mistake somewhere.

But if some transfinite element slipped into the proof, then this is not quite so certain. The contradiction could be the direct result of the use of transfinite elements.

The end of the proof is: $1 = 2$. In this there is no trace of any transfinite ideas. If such ideas nevertheless came into the proof, then the only thing that could have happened, according to the inveterate habit of ideal elements, is that they appeared, did something and disappeared again. Could it be that the proof could be completed without them as some trigonometrical formulae can be proved with the aid of 'i' but also without it?

If only one dangerous element turns up, or if only a few turn up independently of each other, this is in fact the case. Hilbert showed that such proofs could be transformed into harmless proofs, and in these the flaw can readily be found.

Unfortunately ideal elements, like the disembodied ghosts of our imagination which can pass through each other, can turn up in most complicated relationships to each other. And transfinite elements cannot so easily be eliminated from such complicated proofs.

Gentzen noticed that the actual complications turning up in the proofs reminded him of the various complicated ways of rearranging the sequence of natural numbers. If we apply Hilbert's method to such a complicated proof, the transfinite elements do not disappear from it, but the proof is turned into a kind of deduction whose type of complication is similar to some less involved rearrangement of the sequence of numbers. The same thing happens if we repeat Hilbert's procedure, applying it to this less complicated proof. We already know that by progressing through less and less complicated ways of re-arranging the sequence of natural numbers we arrive at a sequence without any complications at all after a finite number of steps. Therefore, using Hilbert's procedure a finite number of times, we should arrive at a proof without any complications at all, i.e. at a proof which does not contain any transfinite elements, and in such a proof the flaw can readily be found.

This is a beautiful and absolutely pure mathematical argu-ment; the result, too, has enormous significance. Our con-fidence in the old procedures can now be restored, at least in the case of the theory of numbers. The majority of mathe-maticians, i.e. those that do not even want to hear about the dangers, still consider the theory of proof as something foreign to them; they think of it as philosophy rather than Mathematics. They recognize the *raison d'être* of a new branch of Mathematics only if it can be made use of creatively in other branches of Mathematics as well. Hilbert wanted to show these people the sort of things that the theory of proof is capable of, and he subjected one of the greatest and best-known problems, the continuum hypothesis of the theory of sets, to the methods of the theory of proof.

The problem is the following: in the sphere of natural numbers, arranged in order of magnitude, there is perfect order. Every number has an immediately succeeding one, for example, 4 comes directly after 3, 13 comes directly after 12. This is utterly impossible in the case of fractions, we can always find other fractions as near as we like to any given fraction. This is even more so if we consider all the real numbers; these stretch right along the line of numbers continuously, quite inextricably coalescing with each other. This is why their totality is usually referred to as the 'continuum'.

We might ask the question about the totalities, introduced by Cantor, whether each number is immediately followed by an-other one. The answer to the question is in the affirmative; from this point of view infinite numbers are similar to natural numbers. The smallest infinite number is that of the natural numbers. We might ask: which is the infinite number that immediately follows it? We know that the continuum, the number of the real numbers, is greater. But is it the one that follows it immediately, or is there perhaps another number between the two? Much research at great depth has been done in connexion with this question; mathematicians are more and more inclined to believe that the continuum is the infinite number that follows immediately the number of the sequence of natural numbers. This is called the 'Continuum hypothesis', or, as those who believe in it very fervently have called it, the 'Continuum theorem'. But no one has got anywhere with it yet.

In quite recent years Gödel (using Hilbert's ideas as his starting point) has proved, using the tools of the theory of proof, that assuming the continuum hypothesis to be true cannot introduce any contradictions into the theory of sets. Therefore the continuum theorem is either independent of the axioms of the theory of sets, or it can be deduced from them. In either case, we can justifiably make use of it in our proofs; it cannot give rise to any contradictions. The proof is similar to the proof of the freedom from contradiction of Bolyai's Geometry; Gödel has constructed a model within the theory of sets, in which both the axioms of the theory of sets and the continuum theorem can get along very well side by side.

After this Hilbert was quite justified in saying to those that still had their doubts about the theory of proof: 'By their fruits ye shall know them.'

Postscript on perception projected to infinity

The freedom from contradiction of the theory of natural numbers is now assured, and the proof can easily be modified to prove the freedom from contradiction of other countable sets, i.e. of the set of positive and negative whole numbers, of the set of fractions or more generally of the set of rational numbers.

We are still left with the set of real numbers, and here we come across new difficulties.

We caught the irrational numbers by means of better and better approximations, by shutting them up in smaller and smaller boxes. So here we are dealing not only with the theory of numbers but with analysis. Here infinite processes occur to right and left all the time, and this brings in new types of danger.

When we were first speaking about this sphere of ideas I was very careful to enunciate quite honestly the very dangerous sentence on which the success or failure of analysis entirely depends. The sentence was this: '*Our perception tells us* that even if we go on indefinitely with the construction of intervals each encased in the previous one, the bit to which they all eventually shrink is a common part of them all.' How can our perception say anything about an infinite process? Have we perhaps forgotten that we have absolutely no right to apply our experiences of the finite to the infinite? We might as well consider another example, which could make us have second thoughts about the matter.

There is no need to be a mathematician to realize that the shortest distance between two points is a straight line. If somebody flies from London to Birmingham, he will get there sooner than if he makes a detour via Bristol:

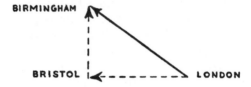

We can see from this immediately that two sides of a triangle together are longer than the third side.

I am nevertheless going to show that the two sides of a right-angled triangle are together exactly equal to the length of the hypotenuse. This is patently stupid, but it is the kind of thing that perception applied to infinite processes is capable of.

Let us draw two steps on the hypotenuse, so that their bounding lines are parallel to the horizontal and the vertical sides respectively:

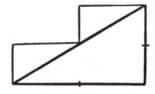

It is obvious that the two vertical pieces on the steps, taken together, are just as long as the vertical side of the triangle, the two horizontal pieces taken together are just as long as the horizontal side. The total length of all the lines drawn to represent the steps is equal to the sum of the lengths of the vertical and horizontal sides.

The same is the case if we draw four steps

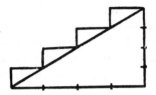

The horizontal pieces taken together are equal to the horizontal side of the triangle, the vertical pieces are equal to the vertical side.

And if we go on subdividing the hypotenuse

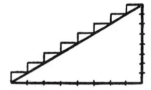 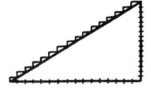

it remains true all the time that the length of the stairway is equal to the lengths of the vertical and the horizontal sides together. On the other hand, as we go on, it becomes less and less possible to distinguish the stairway from the hypotenuse, and 'our perception tells us' that if we go on with the subdivision indefinitely, the stairway will coalesce with the hypotenuse. Accordingly the hypotenuse must be equal to the sum of the other two sides.

After this we might have further thoughts about the reliability of our perception when it is projected into infinity.

It is nevertheless true that the success or failure of analysis depends entirely on that critical sentence. Either we believe it without any basis for doing so, merely because we should like to believe it, or else there is nothing for it but to turn to the methods of the theory of proof. It will then be necessary to consider whether such a statement can lead to a contradiction.

There are therefore further transfinite elements intruding into the system of axioms of analysis. If we admit these, the system will be so wide that not only will the case of transfinite induction used by Gentzen be included in it, but other and much more complicated cases of it as well. Gödel's theorem is still true in this case: the freedom from contradiction of the system cannot be proved by methods which can themselves be formalized within the system considered. There can be therefore absolutely no hope that the methods used up to now will be sufficient to prove the freedom from contradiction of analysis. Here we must look for new methods, probably finer methods than before. This is still an open question today for further research to tackle.

22. *What is Mathematics not capable of?*

THE proof of the freedom from contradiction of the theory of numbers has shown up one of the imperfections of axiomization. The transfinite induction employed there can be expressed in the language of natural numbers, it is a procedure readily conceivable by a finite mind. Nevertheless it has slipped out of the framework of the system of axioms covering the theory of natural numbers.

This is no unique phenomenon. There is no system of axioms that can grasp quite tightly exactly what it is intended to circumscribe. There will always be things that slip through the net, and other things that turn up uninvited. Axiom systems all grasp a lot, and yet they catch but a little.

The fact that axiom systems grasp a lot was shown by the Norwegian mathematician Skolem.

If we want only to catch the sequence of natural numbers with our axioms, i.e. the natural numbers in their original order, some more complicated arrangements of this sequence slip in uninvited, whether we like it or not. It is impossible to separate them from these.

On the other hand if we wish to circumscribe exactly by means of axioms a universe of discourse of greater number than the countable, for instance the set of real numbers, then there will always exist a countable set which somehow finds its way in, which satisfies all the conditions represented by the axioms.

It was Gödel's surprising discovery, namely the fact that any axiom-system worthy of the name which covers the theory of numbers includes problems which are undecidable, which demonstrated that axiom-systems are capable of catching only a little.

Let us consider the exact meaning of the latter statement.

There are plenty of problems in mathematics that up to now have not been decided. I have already mentioned a number of these. For example: are there an infinite number of 'twin' prime numbers? (as for instance 11 and 13, or 29 and 31).

The Goldbach conjecture is likewise undecided. It has been noticed that

$$4 = 2 + 2$$
$$6 = 3 + 3$$
$$8 = 3 + 5$$
$$10 = 3 + 7 = 5 + 5$$

. . . .

i.e. it appears that even numbers greater than 2 can be expressed as sums of prime numbers, sometimes in more ways than one. This is true for the largest numbers that have been examined, but even today it is still a conjecture whether it is true of all even numbers.

Fermat's conjecture is the one that has acquired the greatest fame. We know that

$$3^2 + 4^2 = 9 + 16 = 25 = 5^2$$

i.e. $$3^2 + 4^2 = 5^2$$

and there are other whole numbers besides these such that if we square two of them and add the squares we obtain the square of the third one as a result. Fermat, scribbling notes in the margin of a book, made the remark that he had found a proof that this was impossible for exponents greater than 2, only there was no room for the proof on the margin. The statement in other words is that it is impossible to think of three whole numbers, X, Y, and Z for which the relationship

$$X^3 + Y^3 = Z^3$$
or $$X^4 + Y^4 = Z^4$$
or $$X^5 + Y^5 = Z^5$$

. . . .

would be the case.

Fermat has been dead for some time, and mathematicians ever since have tried to reconstruct his proof, but right up to this day not one of them has met with success. This lack of success, in a case where somebody is reputed to have held the proof in his hands, has aroused such interest about this intrinsically rather uninteresting problem that there have even been some wills in which substantial legacies have been bequeathed to anyone who decides the issue. It is small wonder that this has aroused the imagination of non-professionals even more than the squaring of the circle. Fortunately their ardour has been somewhat damped now that the money bequeathed has lost all its value.

This problem has nevertheless had a fertilizing effect on Mathematics. It has led to the introduction of more ideal elements, the so-called 'ideals', in order to get at the problem, and these have proved very useful in the more important branches of Algebra as well. But, even so, Fermat's conjecture has so far been proved only in the case of particular exponents; in its generality it is still undecided even today. Fermat probably made a mistake, he too probably found a proof only for some special case.

But apart from these there are in Mathematics problems which have been proved insoluble by means of certain circumscribed methods. These are problems which have been decided, but in the negative. Such a problem was, for instance, the solubility of equations of the fifth degree or the effective squaring of the circle. The trisection of an angle and the doubling of the cube also belong to this category of problems. It has been decided that these cannot be effected by means of ruler and compass alone. We can bisect an angle by means of these instruments, but we cannot divide one into three equal parts. The doubling of the cube corresponds to the doubling of our fishpond in three-dimensional space. In the plane we were able to construct the side of the large square with ruler and compass; in three-dimensional space the construction of the edge of a cube whose volume is twice as much as the volume of a given cube is not possible by means of the permitted instruments. This problem is often referred to as the problem of Delos, since the gods apparently required of the people of Delos, who were stricken by plague, to double the size of their altar, which was in the shape of a cube. All the good will in the world was of no avail. Plato later consoled them by telling them that the gods were really using the problem as a means of urging on the Greeks the virtues of studying Geometry.

Gödel's theorem, on the other hand, is not about problems which have up to now not been solved, or decided in the negative, but about problems which are undecid*able* within the relevant system of axioms.

Let us sketch Gödel's argument.

Suppose that we have a well-constructed system of axioms for the science of natural numbers, i.e. for the theory of numbers. In the axioms we have included everything that we are going to need in this field. Of course we have taken care not to bring any contradictions into the system. We have written it all down in the language of symbolic logic, and so every statement assumes the form of a succession of symbols.

Now we can associate a number with every one of these successions of symbols in the same way as we associated pairs of numbers with points in the plane. This may be done in the following way: we have a finite number of mathematical and logical symbols; let us associate the first few prime numbers with these (this time 1 can be included among the prime numbers). For example, let 1 correspond to itself. We shall not need any further symbols for numbers after this, since we can write 2 by writing $1 + 1$, 3 by writing $1 + 1 + 1$ and so on. Let 2 be associated with the symbol '$=$'; let 3 be associated with the symbol meaning 'not', i.e. with '$\sim$'; 5 may be associated with the symbol '$+$' and so on. It makes no difference in what order we do this; let us say that 17 corresponds to the last symbol. Then let the prime numbers beginning with 19 be associated with the letters signifying unknowns like X, Y, . . . since these occur in the statements of the system. For example, 19 could correspond to X, 23 to Y and so on.

In this way we obtain a 'dictionary':

1	.	.	.	.	.	1
$=$	.	.	.	.	.	2
$\sim$	.	.	.	.	.	3
$+$	.	.	.	.	.	5
—	—	—	—	—	—	—
X	.	.	.	.	.	19
Y	.	.	.	.	.	23

We can read from this straight-away that for example the three numbers

$$1, 2, 1$$

correspond to the formula

$$1 = 1$$

Let us make a single number out of the three numbers 1, 2, 1. This can naturally be done quite easily and in many different ways. For example we could add the three numbers, and we should then get 4. The trouble is that this 4 has swallowed the other numbers. It is impossible to see what numbers went to make up this 4, in what order they were, or even how many. For example, 4 might have been

$$1 + 3, \text{ or } 3 + 1, \text{ or } 2 + 2, \text{ or } 1 + 1 + 2, \text{ or } 2 + 1 + 1,$$

certainly not only

$$1 + 2 + 1$$

What we want to do is to construct a number in which we could recognize exactly the parts that went to make it. There is a way of doing this: for example we can multiply the first three prime numbers

$$2, \quad 3, \quad 5$$

each one raised to the power determined by the numbers

$$1, \quad 2, \quad 1$$

In this way we construct the following product:

$$2^1 \times 3^2 \times 5^1 = 10 \times 3^2 = 10 \times 9 = 90$$

So we associate the number

$$90$$

with the formula

$$1 = 1$$

From our number it is easy to recognize the formula with which it is associated. All we have to do is to split it into its prime factors in order of magnitude:

$$
\begin{aligned}
90 &= 2 \times 45 \\
&= 2 \times 3 \times 15 \\
&= 2 \times 3 \times 3 \times 5 \\
&= 2^1 \times 3^2 \times 5^1
\end{aligned}
$$

and so the prime numbers

$$1, \ 2, \ 1$$

have again appeared as exponents, and with these are associated the symbols

$$1, \ = \ , 1$$

in the 'dictionary', so from the number 90 we can correctly write down the formula associated with it, namely the formula

$$1 \ = \ 1$$

With every statement in the system, then, is associated a number. Similarly with every proof we can associate a

number.   A proof is, looked at formally, nothing more than a
succession of statements (in which each statement follows from
the preceding ones).   Numbers have been associated with
statements, and so, if for example a proof consists of three
statements, then three numbers will correspond to it.   These
three numbers can be made into a single number by using the
previous method, and it will always be possible to discern the
constituent part of this number; all we have to do is to split it
into its prime factors.

Supposing that we know that some very large number has
already occurred among the list of associated numbers, and
further that we have the patience of Job to split it into prime
factors, giving say:

$$2^{90\ 000\ 000\ 000\ 000\ 000\ 000} \times 3^{90}$$

Firstly we can see that the exponents are not prime numbers,
and so these do not correspond to a simple statement but to a
proof.   The proof consists of two statements, namely the statements whose associated numbers are the numbers

$$90\ 000\ 000\ 000\ 000\ 000\ 000$$

and

$$90$$

which occur in the exponents.   If we split these two numbers
into their prime factors, we can then reconstruct the statements that correspond to them.   In the first one there are
nineteen zeros, so that this number is

$$9 \times 10^{19} = 3^2 \times 10^{19} = 3^2 \times 2^{19} \times 5^{19}$$

since $10 = 2 \times 5$.   Arranging the bases according to size, we
have

$$2^{19} \times 3^2 \times 5^{19}$$

so the numbers occurring in the exponents are

$$19, \quad 2, \quad 19$$

The factors of the second number are already known to us,

$$90 = 2^1 \times 3^2 \times 5^1$$

so this was constructed out of the three numbers

$$1, \quad 2, \quad 1$$

The 'dictionary' is set down again for convenience:

1	1	We can read out of it that the
=	2	first three numbers,
~	3	i.e. 19, 2, 19
+	5	correspond to the formula
— — — — — — — —		$X = X$
X	19	and the second three numbers,
Y	23	i.e. 1, 2, 1
— — — — — — — —		correspond to the formula
		$1 = 1$

Then what the proof told us is:
If for any arbitrary X we have

$$X = X$$

it follows that

$$1 = 1$$

This is really a miserable little proof, while the number associated with it was of astronomical size. We can imagine how enormous a number might be associated with a proof of some consequence. The essential point, however, is that we know that there is a certain definite number associated with it, and from this number the proof can be reconstructed (not within a lifetime, but at least in principle).

This is one way in which the formulae of a system can be translated into certain natural numbers. But what is the use of it all?

Metamathematics examines the system from the outside; its statements are about formulae or proofs of such-and-such forms in the system. Now these statements can be transformed with the aid of our 'dictionary' so that they concern natural numbers with such-and-such prime factors.

For example while metamathematics is busy examining the formulae of the system which are expressible in terms of the symbols of the system it takes note that the successions of symbols

$$1 = 1$$

and

$$\sim (1 = 1)$$

must be dealt with rather gingerly, since one is the negation of the other. We have already seen that to

$$1 = 1$$

corresponds $$2^1 \times 3^2 \times 5^1 = 90$$

According to the 'dictionary' (apart from the brackets themselves being symbols, we ought really to associate some numbers with these too)

1	1	the sequence	
=	2		3, 1, 2, 1
~	3	corresponds to the formula	
+	5		$\sim (1 = 1)$

and

$$2, \quad 3, \quad 5, \quad 7$$

being the first four prime numbers, the number corresponding to this formula is the number

$$2^3 \times 3^1 \times 5^2 \times 7^1$$

Let us work out this number.

$$2^3 \times 3^1 \times 5^2 \times 7^1 = 2 \times \underbrace{2 \times 2 \times 3 \times 5} \times 5 \times 7$$

$$= 10 \times 10 \times 2 \times 3 \times 7 = 100 \times 42 = 4200$$

Let us put the decompositions into prime factors next to each other:

$$90 = \qquad 2^1 \times 3^2 \times 5^1$$
$$4200 = 2^3 \times 3^1 \times 5^2 \times 7^1$$

Therefore we can reformulate the metamathematical statement: 'The successions of symbols of the forms

$$1 = 1 \text{ and } \sim (1 = 1)$$

express the contraries of each other' in the following way:

'90 and 4200 are numbers which are such that the decomposition into prime factors of the latter begins with 2^3, and the exponents of the prime numbers coming after this are the same as the exponents figuring in the prime decomposition of 90.'

In the last sentence there is no trace of metamathematics, it is a statement about numbers. The system considered has the avowed purpose of formulating statements about numbers.

Thus this sentence can also be written down by means of the symbols of the system considered, so that not a single word will remain in it. It will become one of the ordinary grey successions of symbols; it is not immediately obvious that it has two interpretations. But it has, in fact, two different interpretations; it can be read as two different texts. One is a text which says something about numbers, which can be obtained from any formula of the system, if we remember the original contents of the symbols. The other text is the metamathematical statement which it embodies.

As Gödel was playing about with such successions of symbols that had two interpretations, he came across a number, say 8 billion. We know exactly how it is built up out of its prime factors, but a whole lifetime would not be long enough for its actual computation. Gödel noticed that this number is able to give the following information. If, using the symbols of the system in the way we did in the case of the sentence just discussed, we write down the mathematical statement:

'The formula corresponding to 8 billion is not provable in the system'

and inquire for the number which corresponds to the formula thus obtained according to the dictionary, we shall find to our amazement that this number is just 8 billion. So 'the formula corresponding to 8 billion' is the same formula. So the statement states in one of its senses:

'I am not provable'

Let us make it quite clear that this is no playing with words, nor any kind of sophistry. There is an ordinary grey formula in front of us, undeniably a succession of symbols, just like the others. It is only when we see, with the aid of our 'dictionary', the double meaning which has been smuggled into this succession of symbols by metamathematics that we notice that it is humming the following tune, looking all innocent:

'I am not provable'

It is small wonder that this formula is undecidable within the system, in spite of expressing an innocent number theoretical statement through its other meaning.

If it could be proved, then it would be in contradiction to what its metamathematical meaning asserts, namely just that it is not provable.

If, on the other hand, it could be disproved, then it would be this disproof which would establish the metamathematical statement contained in it. So its disproof would be its proof.

So it is impossible to prove it or to disprove it. It is undecidable.

It must be emphasized again that if we did not remember the 'dictionary' this would be an ordinary grey formula of the system, some quite innocent number theoretical statement about additions and multiplications. Gödel proved the existence of undecidable formulae of this kind in every system. It is not impossible that the Goldbach conjecture may be among them. Quite possibly the reason why it has not been found possible to decide the issue is that if we set up a system of axioms out of all the tools used by those who have attempted to settle this conjecture, it may happen that through the 'dictionary' it would just be humming:

'I am not provable within the system'

The same applies to any other problem which has not yet been solved. Every mathematician must look this possibility in the face.

There might be one more objection to all this. The trouble might be due to imperfection in axiom systems. Surely even such 'Gödel problems' can be decided if we do not restrict ourselves to any particular axiom-system. But now Church has constructed a problem which is not decidable by means of any of the arguments that mathematicians today can think of, quite independently of whether these arguments can be circumscribed by means of any axiom-system.

This is where I must stop writing. We have come up against the limits of present-day mathematical thinking. Our epoch is the epoch of increasing consciousness; in this field Mathematics has done its bit. It has made us conscious of the limits of its own capabilities.

But have we come up against final obstacles? Up to the present there has always been a way out of all the *culs-de-sac* encountered in the history of mathematics. There is one point about Church's proof which we might do well to ponder over: it would be necessary to formulate quite precisely what the arguments are that mathematicians today can think of, if we wanted to employ the processes of Mathematics in connexion with such a concept. The moment something is formulated, it is already circumscribed. Every fence encloses a narrow space. The undecidable problems that turn up manage to get through the fence.

Future development is sure to enlarge the framework, even if we cannot as yet see how. The eternal lesson is that Mathematics is not something static, closed, but living and developing. Try as we may to constrain it into a closed form, it finds an outlet somewhere and escapes alive.

After Use

Should the reader wish to refer to, for instance, the integral he will find in the list of Contents only the title 'many small make a great'. For this reason I have added here as a postscript what mathematical concepts are to be found in the various chapters. (Don't be put off by them!)

Part III

A CATALOGUE OF SELECTED DOVER BOOKS
IN ALL FIELDS OF INTEREST

A CATALOGUE OF SELECTED DOVER BOOKS
IN ALL FIELDS OF INTEREST

AMERICA'S OLD MASTERS, James T. Flexner. Four men emerged unexpectedly from provincial 18th century America to leadership in European art: Benjamin West, J. S. Copley, C. R. Peale, Gilbert Stuart. Brilliant coverage of lives and contributions. Revised, 1967 edition. 69 plates. 365pp. of text.

21806-6 Paperbound $3.00

FIRST FLOWERS OF OUR WILDERNESS: AMERICAN PAINTING, THE COLONIAL PERIOD, James T. Flexner. Painters, and regional painting traditions from earliest Colonial times up to the emergence of Copley, West and Peale Sr., Foster, Gustavus Hesselius, Feke, John Smibert and many anonymous painters in the primitive manner. Engaging presentation, with 162 illustrations. xxii + 368pp.

22180-6 Paperbound $3.50

THE LIGHT OF DISTANT SKIES: AMERICAN PAINTING, 1760-1835, James T. Flexner. The great generation of early American painters goes to Europe to learn and to teach: West, Copley, Gilbert Stuart and others. Allston, Trumbull, Morse; also contemporary American painters—primitives, derivatives, academics—who remained in America. 102 illustrations. xiii + 306pp. 22179-2 Paperbound $3.50

A HISTORY OF THE RISE AND PROGRESS OF THE ARTS OF DESIGN IN THE UNITED STATES, William Dunlap. Much the richest mine of information on early American painters, sculptors, architects, engravers, miniaturists, etc. The only source of information for scores of artists, the major primary source for many others. Unabridged reprint of rare original 1834 edition, with new introduction by James T. Flexner, and 394 new illustrations. Edited by Rita Weiss. 6⅝ x 9⅝.

21695-0, 21696-9, 21697-7 Three volumes, Paperbound $15.00

EPOCHS OF CHINESE AND JAPANESE ART, Ernest F. Fenollosa. From primitive Chinese art to the 20th century, thorough history, explanation of every important art period and form, including Japanese woodcuts; main stress on China and Japan, but Tibet, Korea also included. Still unexcelled for its detailed, rich coverage of cultural background, aesthetic elements, diffusion studies, particularly of the historical period. 2nd, 1913 edition. 242 illustrations. lii + 439pp. of text.

20364-6, 20365-4 Two volumes, Paperbound $6.00

THE GENTLE ART OF MAKING ENEMIES, James A. M. Whistler. Greatest wit of his day deflates Oscar Wilde, Ruskin, Swinburne; strikes back at inane critics, exhibitions, art journalism; aesthetics of impressionist revolution in most striking form. Highly readable classic by great painter. Reproduction of edition designed by Whistler. Introduction by Alfred Werner. xxxvi + 334pp.

21875-9 Paperbound $3.00

MATHEMATICAL PUZZLES FOR BEGINNERS AND ENTHUSIASTS, Geoffrey Mott-Smith. 189 puzzles from easy to difficult—involving arithmetic, logic, algebra, properties of digits, probability, etc.—for enjoyment and mental stimulus. Explanation of mathematical principles behind the puzzles. 135 illustrations. viii + 248pp.
20198-8 Paperbound $2.00

PAPER FOLDING FOR BEGINNERS, William D. Murray and Francis J. Rigney. Easiest book on the market, clearest instructions on making interesting, beautiful origami. Sail boats, cups, roosters, frogs that move legs, bonbon boxes, standing birds, etc. 40 projects; more than 275 diagrams and photographs. 94pp.
20713-7 Paperbound $1.00

TRICKS AND GAMES ON THE POOL TABLE, Fred Herrmann. 79 tricks and games— some solitaires, some for two or more players, some competitive games—to entertain you between formal games. Mystifying shots and throws, unusual caroms, tricks involving such props as cork, coins, a hat, etc. Formerly *Fun on the Pool Table*. 77 figures. 95pp.
21814-7 Paperbound $1.25

HAND SHADOWS TO BE THROWN UPON THE WALL: A SERIES OF NOVEL AND AMUSING FIGURES FORMED BY THE HAND, Henry Bursill. Delightful picturebook from great-grandfather's day shows how to make 18 different hand shadows: a bird that flies, duck that quacks, dog that wags his tail, camel, goose, deer, boy, turtle, etc. Only book of its sort. vi + 33pp. 6½ x 9¼. 21779-5 Paperbound $1.00

WHITTLING AND WOODCARVING, E. J. Tangerman. 18th printing of best book on market. "If you can cut a potato you can carve" toys and puzzles, chains, chessmen, caricatures, masks, frames, woodcut blocks, surface patterns, much more. Information on tools, woods, techniques. Also goes into serious wood sculpture from Middle Ages to present, East and West. 464 photos, figures. x + 293pp.
20965-2 Paperbound $2.50

HISTORY OF PHILOSOPHY, Julián Marias. Possibly the clearest, most easily followed, best planned, most useful one-volume history of philosophy on the market; neither skimpy nor overfull. Full details on system of every major philosopher and dozens of less important thinkers from pre-Socratics up to Existentialism and later. Strong on many European figures usually omitted. Has gone through dozens of editions in Europe. 1966 edition, translated by Stanley Appelbaum and Clarence Strowbridge. xviii + 505pp. 21739-6 Paperbound $3.50

YOGA: A SCIENTIFIC EVALUATION, Kovoor T. Behanan. Scientific but non-technical study of physiological results of yoga exercises; done under auspices of Yale U. Relations to Indian thought, to psychoanalysis, etc. 16 photos. xxiii + 270pp.
20505-3 Paperbound $2.50

Prices subject to change without notice.
Available at your book dealer or write for free catalogue to Dept. GI, Dover Publications, Inc., 180 Varick St., N. Y., N. Y. 10014. Dover publishes more than 150 books each year on science, elementary and advanced mathematics, biology, music, art, literary history, social sciences and other areas.